The publisher gratefully acknowledges the generous support of the August and Susan Frugé Endowment Fund in California Natural History of the University of California Press Foundation.

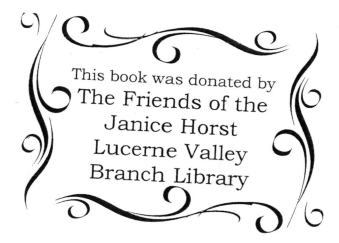

Chuckwalla Land

Chuckwalla Land

THE RIDDLE OF CALIFORNIA'S DESERT

DAVID RAINS WALLACE

UNIVERSITY OF CALIFORNIA PRESS
Berkeley   Los Angeles   London

University of California Press, one of the most distinguished university presses in the United States, enriches lives around the world by advancing scholarship in the humanities, social sciences, and natural sciences. Its activities are supported by the UC Press Foundation and by philanthropic contributions from individuals and institutions. For more information, visit www.ucpress.edu.

University of California Press
Berkeley and Los Angeles, California

University of California Press, Ltd.
London, England

Library of Congress Cataloging-in-Publication Data

Wallace, David Rains, 1945–.
  Chuckwalla land : the riddle of California's desert / David Rains Wallace.
    p. cm.
  Includes bibliographical references and index.
  ISBN 978-0-520-25616-3 (cloth, alk. paper)
    1. Desert biology—California.   2. Deserts—California.   I. Title.
QH105.C2W338   2011
578.75409794—dc22   2010029593

Manufactured in the United States of America

19 18 17 16 15 14 13 12 11
10 9 8 7 6 5 4 3 2 1

This book is printed on Cascades Enviro 100, a 100% post consumer waste, recycled, de-inked fiber. FSC recycled certified and processed chlorine free. It is acid free, Ecologo certified, and manufactured by BioGas energy.

To Mike Kowalewski

Arcadia (also Arcady). From *Arcadia*, pastoral region of ancient Greece regarded as a rural paradise . . .: a usually idealized region or scene.

Webster's Third New International Dictionary

A shape with lion body and the head of a man,
A gaze blank and pitiless as the sun,
Is moving its slow thighs, while all about it
Reel shadows of the indignant desert birds.

W.B. Yeats, *The Second Coming*

Contents

Prologue

BUSHES AND LIZARDS

It took me years to notice the California desert. When I first crossed it, on freeways from the east, it seemed more of the same blazing scrub as in Nevada or Arizona. When I crossed it from the west, it seemed more of the same agroindustrial sprawl that borders California freeways. It wasn't all subdivisions, warehouses, tomato fields and power lines, not yet, but it looked more like an enormous vacant lot than a landscape. Tractmongers' catchphrases—"raw land," "nothing there"—nonsensical applied to forests or wetlands, sounded more appropriate to the dead-looking brush sliding past the car windows.

I read Mary Austin's *Land of Little Rain* and admired the stubborn sensibility of her explorations: "Go as far as you dare in the heart of a lonely land, you cannot go so far that life and death are not before you . . . out of the stark, treeless waste rings the music of the night-singing mockingbird." But her desert resembled her 1903 prose, sepia toned, like photographs of twenty-mule teams in county museums. I went to Nevada or Arizona if I wanted up-to-date desert, and even that felt secondhand—if not Mary Austin sepia, then Edward Abbey Kodachrome: "Death and life usually appear close together, sometimes side by side, in the desert. Perhaps that is the secret of the desert's fascination. . . . Nothing, not even the waiting vulture in the sky, looks more deathly than a dying giant cactus."

I didn't really look at California desert until I had to write something about it in 1983, and even then I planned it as a diversion. I was more interested in Central Valley riparian woodland. Since little of that remained in the Central Valley, I decided to visit a Nature Conservancy preserve on the Kern River just west of Walker Pass, one of the historic gateways to the Mojave Desert. I imagined spending an afternoon in a cathedral of giant oaks, walnuts, sycamores, and box elders such as John Muir had described, and then paying a brisk duty call on the Mary Austin country.

The Conservancy preserve failed romantic expectations. Instead of hardwood gallery forest in a stately valley, I found a large willow and cottonwood thicket hugging a rocky canyon bottom, good habitat for rare yellow-billed cuckoos, less so for a contemplative afternoon. After a chat with the preserve manager and a dispirited stroll past the thicket, I got back in the car and headed east without much anticipation. But the desert had surprises for me.

To start with, it refused to wait for me across Walker Pass. Soon after I pulled away from the willow thicket, troops of olive drab spiky plants began clustering around irrigated pastures—Joshua trees—although grassland and oaks still covered the canyon sides. The tall yuccas seemed oddly zoomorphic, almost to be moving west, an impression abetted by my eastward momentum. They had an animation that I hadn't associated with desert, which made me wonder exactly what desert is aside from an assemblage of unsightly (or intriguing, depending on the viewpoint) organisms in places too dry for normal ones.

Driving on a two-lane road instead of a freeway contributed to my new curiosity. After crossing the pass and winding into the sepia jumble beyond, I kept passing spots that looked interesting. I eventually found one where I could pull over, a dirt parking lot at Red Rock Canyon State Park, which, according to a bullet-riddled sign, was a famous place where many Hollywood Westerns had been filmed. I'd never heard of it.

The lot wasn't encouraging, a litter of glass, paper, Styrofoam, and worse. Someone with a bleeding ulcer had been sick under a dying yucca. I fled into a gulch, not expecting much. The parking lot's squalor was hard to escape. Trash and ORV tracks dogged me, the tracks running impartially over bare rock, deep sand, and surprisingly steep and narrow passages. "Go as far as you dare in the heart of a lonely land," Mary Austin might have written in 1983, "you cannot go so far that ORV tracks are not before you."

But it was April and I started seeing wildflowers, first a clump of yellow monkey flowers at a seep, then scatterings of tiny goldfields, showy deep violet gilias, big pale violet evening primroses, small white and brick red evening primroses, pale violet larkspurs, royal blue lupines, surprisingly lush green desert rhubarbs, scarlet paintbrushes. The flowers didn't carpet the ground as in magazine features, but their sparsity made them all the more impressive. Each inflorescence seemed to leap from the gray sand, mimicking the Walker Pass Joshua trees' apparent animation. Even the dead-stick creosote bushes scattered over the flats and benches had little bell-like yellow flowers teeming with bees.

I followed the gulch past dark red walls of volcanic ash studded with black crystals. The only sound once I'd turned a few corners was the wind that had pursued me from the pass. The only motions were the flowers shaking in the wind and a prairie falcon that dropped from a ledge and glided past my head. After a while, the walls narrowed between two huge boulders shaped like deformed human skulls, one red and one yellow. The red one, flattened and elongated, leaned forward as though to peer into the canyon floor; the yellow one, beetling and bulbous, tilted up at the sky. Past them, the badland formations, or "hoodoos," grew even more bizarre. Gray green tuff sprouted fungoid and phallic shapes; red ash erupted spires and gothic facades. In places, the facades had collapsed into alcoves so palatial looking that I half expected to see ruined tiles and fountains among the sand and weeds.

The grotesquerie was unexpectedly enchanting. The weird formations seemed to act directly on my mind, to knead and stretch it, squeezing aside the worn furniture of normal perception. Two black army helicopters that thundered overhead failed to dispel the strangeness; in fact, they fit right in, more like mutant dragonflies from an atomic hinterland than enforcers of agroindustrial growth. Even when the canyon widened into a riparian gallery where mockingbirds sang and rabbits hopped, the place kept an air of "through the looking glass." A pair of Say's phoebes copulated on a stubby phallus of tuff as though to burlesque suburban domesticity.

As dusk fell and a gigantesque full moon loomed in the east, I turned back and descended a gulch I barely recognized, its red and yellow gargoyles changed into blue and gray trolls. The floor glowed in the halflight, and the sky was a lambent indigo under which the flowers seemed not just to leap from the sand but to incandesce. Above the parking lot, golden eagles carried surprisingly long sticks to a nest cliff against the dusty orange horizon.

The desert resumed a vacant expression as I followed the freeways beyond Barstow the next day. But when I pulled into a campground at Mitchell Caverns State Park in the eastern Mojave, it came to life again. A roadrunner lunged from a spiky shrub to peck at insects caught on my radiator grill. Another ran up to take a look, and then both returned to the shrub, where I supposed they were nesting in proximity to fast food as cars pulled in and out of the lot. It was an aggressive reception from what had seemed a passive continuum.

Don MacNeil, a curator at the Oakland Museum, had said that if I wanted to write about the California desert, Mitchell Caverns was the place to go. As I walked around there and in the surrounding Providence Mountains during the next few days, I got impressions different from any I had experienced in desert before.

The lizards struck me first. There were so many, and so many kinds. Some were ordinary-looking species such as scuttle about a lot of places—

little side-blotched lizards, medium-size whiptails and spiny lizards. Even they were unusually lively at Mitchell Caverns, as were stranger kinds. In childhood, I had possessed a small, pale horned lizard discouragingly lacking in vitality, not surprisingly, since I hadn't known how to feed it. A much larger, vividly marked individual in the park was another matter. Presiding stoutly over an anthill, it fixed me with an obsidian gaze and disdained to flee, obdurate as a porcupine.

Some kinds seemed to transcend "lizardness" as I knew it. In dry washes, zebra-tailed lizards whipped across my path every few yards, curling banded tails over their backs as though ready to lash out scorpion-like at a pursuer. They ran with bodies lifted fully off the ground by legs that moved so fast they seemed to revolve like a windup toy. On the flats, shady bushes often produced big, tan leopard lizards that got up and darted off on their hind legs like minidinosaurs. Running farther and faster than I'd thought lizards could, these species seemed a contravention of sprawling reptilian normality, as though impelled by an unexpected, supersaurian energy.

Other kinds were more sedate but even more substantial. Desert iguanas with dragonlike crested spines eyed me from the flats, or, once, from the branches of a creosote bush. In foothills, fat blackish chuckwallas sidled about on boulders with a kind of crumpled gravity, like lizard cows. And that, in fact, is what desert iguanas and chuckwallas are—herbivores, primary consumers—unusual niches for latter-day saurians. These two species get not only their food from plants but their water. According to Raymond B. Cowles, a UCLA biologist who studied desert lizards for fifty years, captive desert iguanas and chuckwallas can die of dehydration beside untouched water pans, although they can be taught to drink: "The best way of inducing these novices to consume water is to allow drops to splash into a Petri dish or other shallow receptacle to create ripples. Apparently the stimulus of motion leads them to water. They then usually plunge their noses into the water and discover for themselves the delights of satisfying their growing thirst."

In coastal woodland I was used to seeing fence lizards, alligator lizards, and skinks, but they seemed peripheral because they are opportunistic omnivores, snatching bugs at the margins of the starring herbivore-carnivore roles that coastal mammals and birds play. Of course, many mammals and birds inhabit the Providence Mountains. Desert wood rat houses built of cactus were as numerous as the stick houses of dusky-footed wood rats in coastal woodland. Kangaroo rat burrows were so thick that I sank into them as I walked. But the only visible mammals in the daytime heat were occasional ground squirrels and jackrabbits. Even the abundant black-chinned sparrows, phainopeplas, Gambel's quail, and mourning doves largely vanished at noon. The lizards, especially the big herbivores and carnivores, seemed the main act, which hinted at an alternative to the received wisdom that evolution has left the lowly reptiles behind.

The hint gave me a renewed interest in the shrubs that looked dead from the freeway. They'd appeared an unlikely base for such a pyramid of life, but there they were. To be sure, showy herbaceous wildflowers bloomed all over the Providence Mountains in that season. Each creosote bush had collected a bouquet of purplish white phacelias at its base, grown from wind-deposited seeds. Along with these, the lizards and desert tortoises I encountered were eating desert dandelions, desert chickory, desert pincushion, fiddlenecks, and gold poppies. But the herbs would wither in a few weeks, and the bovine reptiles would go back to browsing the inedible-looking scrub, at which I began to look more closely as I idled around the flats, dry washes, and canyons.

At first glance, the scrub seemed comprised of two species, the rangy creosote bush and a squat, whitish shrub called burroweed. They dotted the brown land in all directions, a situation that prevails throughout California's Mojave and Sonoran deserts and encourages the freeway-view impression of moribund monotony. Nobody seems to know or care much why these two associate so stubbornly. They have little in common beside a penchant for the driest desert. The roots of both exude chemicals that inhibit the growth of other plants, including their own species. But they are always together, like endlessly replicated vegetable versions of Don Quixote and Sancho Panza.

Creosote bush has an aloof, shabby-genteel air with its tall stems shooting from bare gravel. Its golden spring flowering can be posh in good years, although its other features are not—tiny, paired leaflets that emerge green after rains but soon turn olive drab, hard brown fruits, and the tarry-smelling resin for which it is named. It is an errant black sheep of an aristocratic family, the Zygophyllaceae, which boasts one of the world's noblest trees, the genus *Guaiacum*, called lignum vitae, the "wood of life," because it is full of precious oil. *Guaiacum* is so valuable that it is endangered in its tropical forest habitat. Most of creosote bush's other relatives live in the tropics, and its presence in California desert is something of a mystery. Botanists aren't even sure—depending on its obscure relationships with South American congeners—whether to call it *Larrea divaricata* or *L. tridentata.*

Burroweed, also called burrobush or white bur sage, is distinctly plebeian, a member of the teeming sunflower family and a close relative to despised ragweed, with similarly unlovely foliage and hay fever–inducing pollen. Much of the time, burroweed has no foliage, and when the thin, hairy leaves and spiny flowers emerge after rains, it's not that much of a change. The species is named for being too bitter for any livestock except burros to browse. Many other disreputable California species share its genus, which may be *Franseria* or *Ambrosia* depending on the authority.

Another "bush" figured in the apparent monotony, but the spindly, yellowish jumping cholla cactuses that loitered threateningly among the Dons and Sanchos did not seem like shrubs in the same sense as the first two species. Creosote bush and burroweed leaves may be unpalatable for most animals for much of the year, but they *are* leaves, with a faint promise of carbohydrates and proteins. Cholla has only wickedly barbed spines, good for wood rat houses but not much else except, of course, for the cholla itself. Although classed in the same genus, *Opuntia*, as lushly fruiting prickly pear, it can reproduce without seeds because it clones from twigs that—like science-fiction parasites—stick so quickly to passers-by that they seem to jump on them.

Looking closer, however, I began to see surprisingly numerous variations on these three species' themes of spikiness, brittleness, leaflessness, stickiness, grayness, hairiness, and smelliness. First I noticed a lot of ephedras, green-stemmed, broomlike plants whose tiny scalelike needles and cones identify them, unexpectedly, as desert gymnosperms, distant relatives of pines and redwoods. Then I saw that a lot of shrubs that I'd thought were ephedra had tiny purple blossoms on their stems. They were turpentine broom, *Thamnosma,* not a gymnosperm but an angiosperm related to citrus.

The more I looked, the more diversity I saw. Another bush with scalelike leaves and tiny blossoms was blackbush, a yellow-flowered rose relative, as was another called bitterbrush, with yellow flowers so fragrant I could smell it yards away. I saw many pea relatives: indigo bush, with scalelike leaves and tiny blue flowers; honey mesquite, just putting out fernlike compound leaves; catclaw acacia, with bottlebrush-like yellow flowers smelling of baby powder. I saw even more sunflower relatives: cheesebush, like a larger, greener, even more disheveled and smelly (thus the name) version of burroweed; desert fir, with prickly evergreen foliage that did look deceptively coniferous; felt thorn, with fuzzy spikes for leaves; brittlebush, with low leaf clumps brandishing sunflower-like blossoms.

I saw beet relatives—desert hollies, saltbushes, and shadscales. I saw mint relatives—blue sages, paper-bag bushes. I saw lily relatives—banana yuccas, Mojave yuccas. There were shrubs so obscure that they lack familiar common names—*Lycium,* a tomato relative, *Menodora,* an olive relative. California desert has shrubs so obscure that they lack even common relatives. *Krameria* is a spiny little bush with purplish, orchid-like flowers that botanists once classed with peas but have since granted a family all its own. The family has one genus, of which California has two species, both desert dwellers and both semiparasitic, able to tap fluid from other plants with their roots. *Crossosoma* is a spiny medium-size bush with pinkish white flowers that botanists have never linked to another group. Again, the family consists of a single genus with two California species—one in the desert, one on offshore islands.

As with the lizards, this scrubby prodigality hinted at unfamiliar evolutionary tendencies. Of course, many bushes grow in California's nondeserts. They dominate great expanses of coastal and mountain chaparral, and throng woodland understories. Yet, as with the coastal lizards, they are less diverse and idiosyncratic than the desert bushes. Many species belong to a few common genera such as manzanita, ceanothus, rose, and currant. They seem peripheral to the main nondesert theme of trees. Many are simply shrub versions of trees—scrub oak, vine maple, dwarf alder.

I had gotten interested in plant evolution in northwest California, where the subject arises naturally from the lush, venerable forest. Assuming that the first land plants were tiny, 350-foot-tall redwoods clearly have undergone a lot of evolution, and there is plenty of other evidence for this. Redwood fossils occur across the Northern Hemisphere in formations going back to the Jurassic period two hundred million years ago. Fossils show that a forest of great conifers, and later of angiosperm trees too, covered what are now North America and Eurasia for most of that time. The Mojave's bushes made a less obvious case for plant evolution. They didn't look that venerable, in fact, many didn't look that evolved—more like *objets* contrived from coat hangers and long-discarded house plants by a technologically challenged conceptual artist.

I returned from my 1983 trip in a state of exhilarated puzzlement. The presence of so much interesting life in an apparently dead place implied an unexplained ambiguity. The desert was so ethereally surreal at Red Rock Canyon yet so earthily real at Mitchell Caverns. Forming any clear ideas about this seemed unlikely, however, and I didn't think of trying. So much had been written and was being written about both surreal and real desert that I doubted I could add much.

But I kept going back, partly because protecting California desert was a big issue in the late twentieth century so I had more work there, partly just because I'd gotten attached to it, particularly the bushes and lizards. In the process, I gradually realized that there was something that hadn't been addressed much in desert writing, whether surrealist or realist, something at the heart of the desert's ambiguity. I realized that—although there have been many ideas on the subject—nobody really knows how the bushes and lizards got into the desert or how the desert itself got into California. Nobody even knows how old it is.

There is evidence that the present California desert is very recent in geological time. Desert valleys contain beach terraces and mollusk shells showing that huge lakes filled them before about ten thousand years ago. Preserved by dryness, the remains of wood rat nests over nine thousand years old contain largely nondesert plants like juniper and pine, showing that woodlands then grew in places where creosote bush and burroweed prevail now. This recentness is part of a global phenomenon. More rain fell on midlatitudes when the earth was cooler during the last continental glaciations, and woodlands covered present African, Middle Eastern, and Asian deserts as well.

But does this recentness mean that desert itself is recent? From the freeway, it might well seem so. Apparent simplicity and monotony do suggest that only a few hardy plants have been able to adapt to a world of unprecedented dryness. I once heard an old rancher say that we should study desert plant adaptations because the world is drying out as we deplete its water. He saw such study as an alternative to limiting the amount of groundwater he could pump for his cows.

But then, the diversity in places like the Providence Mountains shows that there is more to desert evolution than meets the casual eye. And we have reason to assume that extreme aridity and heat have affected much of the planet for long periods. Desert-forming factors—cooling and drying of the air as it flows from tropical to temperate zones, formation of rain shadows as rising mountains block moist oceanic air from continental interiors—have operated through the roughly four hundred million years of land plant and animal existence. With a long time to evolve, desert plants could be very old.

California desert has characteristics that might suggest a long past. Hints of ancient connections between it and others exist. Although it may look as monotonous from a freeway near the Oregon border as from one near the Mexican border, it has three subdivisions with widely separate geographical affinities. The Great Basin Desert east of the Cascades and Sierra Nevada is a cold desert where severe winters exclude many of the bushes and lizards that I found near Mitchell Caverns. Sagebrush and shadscale, its dominant shrubs, have many relatives in chilly central Asia. The Mojave Desert east of the southern Sierra and Transverse Ranges is a warm desert, although frequent winter frost excludes some species. Its yuccas, creosote bushes, and burroweeds have similarities to plants in the Chihuahuan Desert of Texas and north-central Mexico. The Sonoran Desert east of the San Bernardino Mountains and Peninsular

Ranges and south through most of Baja is a hot desert where infrequency of frost allows subtropical vegetation. Similar desert grows in South America as well as Arizona and northwest Mexico.

Whether long or short, the desert's past remains enigmatic because no known fossil continuum comparable to that of forests or grasslands reaches into it. This may be a part of its nature. Fossils mainly form when dead organisms accumulate in water-deposited sediments. In deserts, wind and flash floods erode away more sediments than sparse waters deposit. Forest fossils going back many millions of years have been found in today's California desert, but few desert fossils. There are desert bush fossils, but not many, and they don't form an unbroken succession going back two hundred million years on this or any other continent.

Enigmas loom larger in California's desert than others because it is North America's most extreme one, its driest and hottest, with conditions that block rainfall from all directions. The Mojave's Death Valley is the archetype of this. "Surely this is North America's most barren desert," wrote a puzzled Edward Abbey of his only stay in California desert. "A dull, monotonous terrain, dun-colored, supporting a few types of shrubs." He was used to the Arizona Sonoran's giant cactuses and small trees, which grow because summer rains from the Gulf of Mexico reach them. But in Alta California, even Sonoran desert gets so little summer rain that saguaros and elephant trees grow only near the Arizona and Baja borders. It has thus been called the Colorado Desert, but it is really just desiccated Sonoran. Much of Baja supports giant cactuses but there the desert comes right to the Pacific, where its location at dead center in the Northern Hemisphere arid belt assures that fog drip is almost the only regular moisture.

The California desert's inscrutable extremes baffled and repelled explorers and early naturalists. In the past century, however, the very proximity to urban explosion that can make it seem insignificant has focused a particular amount of attention on it. Some major scientists have wondered about desert origins here, including one of the twentieth century's great evolutionists, G. Ledyard Stebbins, a botanist and geneticist who helped create the neo-Darwinian synthesis that has been biology's fulcrum since the 1950s. This attention has generated considerable knowledge about evolution in the California desert. It has yet to solve the riddle of how and why the desert originated.

When Stebbins addressed the problem in a pair of influential essays, mostly about the surprising diversity of desert bushes, the response was typically ambiguous. Daniel I. Axelrod, a leading U.S. paleobotanist, thought they supported his idea that the desert is very young. Jerzy Rzedowski, a leading Mexican botanist, thought they supported his idea that the desert is very old. Stebbins, who died in 2000 at the age of ninety-four, seems to have remained unsure about the desert's age.

The desert past remains a prickly hiatus in conventional evolutionary narratives, which flow conveniently from fish and amphibians to birds and mammals, from algae and ferns to trees and grasses. A landscape of spiky shrubs and shrub-eating lizards does not fit easily into those narratives. Whether very new or very old, it implies anomalies, which, even if natural, remain enigmatic. This book won't try to link California desert with conventional narratives. It will describe attempts to make such links, and the enigmas they have evoked.

ONE

A Sphinx in Arcady

The desert's stark reticence challenges comfortable notions that we humans occupy the apex of benign, reasonable processes that have unfolded especially to produce us. American pioneers saw California's forests and grasslands as an unspoiled Promised Land divinely created to usher in an Arcadian Golden Age for progressive civilization. They saw its desert as a strange ruined wasteland that needed "reclamation," with the implication that it somehow is not natural as forests and grasslands are, that it might be part of a contrary, hostile creation.

Early conservationists like John Muir did not necessarily share the public's biblical creationism (Muir had his own pantheist version), but they did share its faith in progress. Enthusiastic about preserving forest, they were less attached to desert and tended to accept the reclamation paradigm. Muir was a farmer, if a reluctant one, and expected that "the fertilizing waters of the rivers" would irrigate much of arid California "giving rise to prosperous towns, wealth, arts, etc." Even a desert enthusiast like Mary Austin put protecting Owens Valley irrigation water from thirsty Los Angeles before protecting desert flora and fauna. Herself a homesteader, she romanticized and empathized with the herders and miners she knew in the 1890s.

Joseph Smeaton Chase, an English writer who explored the desert on horseback during World War I, evoked the prevailing attitude in mythic terms:

> The mountains, the sea, even the vast and changeful sky, have each some predominant genius for those who love the fair features of our earth. What sentiment does the desert yield by which it may be linked with human emotions? What analogy exists by which we may come in touch with it? The answer must be, There is none. At every point the desert meets us with a negative. Like the Sphinx, there is no answer to its riddle. It is in the fascination of the unknowable, in the challenge of some old, unbroken secret, that the charm of the desert consists. And the charm is undying, for the secret is—Secrecy.

Chase's metaphor of the desert as a Sphinx like the colossus before the Giza pyramids, "fixed in eternal reverie amid the immemorial sands of Egypt," might seem a cliché. It did arise partly from the mystique that effloresced in archaeological sensations like King Tut's tomb and in popular spin-offs like *The Mummy,* the 1932 film starring Boris Karloff as an ancient Egyptian priest who is dug up from the desert and resurrected by an ancient spell. Living in Palm Springs when Hollywood was starting to colonize it, Chase drew on and contributed to the mystique in *California Desert Trails,* his 1919 account of his adventures. He published another desert book promoting tourism with the distinctly clichéd title of *Our Araby.*

The Sphinx is a complicated monster, however, with a past almost as mysterious as the desert's. According to mythologist Joseph Campbell, the Egyptian Sphinx was the son of the Creator, Ptah, represented as a shrouded, masked human mummy, and Sekhmet, the lion goddess of the desert sun's destructive potential. The human-headed, lion-bodied monster manifested Pharaoh's power over life and death and thus guarded royal tombs and, by extension, the stabilizing power of divine royalty. He also, according to the *Encyclopedia Britannica*, presided over religious rites that involved riddles.

Around four thousand years ago, "through Egyptian influence" the *Encyclopedia* says, the Sphinx spread into the Levant, ancient Canaan and Phoenicia. There he underwent a sex change, acquiring a female face and breasts, and also wings. The significance of this is unclear, but the transsexual monster evidently retained her power. When she appeared among the emerging cities of Mycenaean Greece some five hundred years later, the Sphinx played a pivotal if enigmatic role in one of western civilization's central myths.

Legend says a Phoenician prince, Cadmus, founded Thebes, the Greek city where the myth takes place. The remains of its early structures do contain Middle Eastern artifacts. In the myth, a Theban king, Laius, orders that his infant son be exposed—hung by his pierced feet from a bush—because of an oracle's prophecy that he will kill his father. But the herdsman assigned to the job passes the infant to nearby Corinth's childless royal family, who adopt him. Named Oedipus, "swollen foot," by his foster parents, the boy grows up to enact an archetype of disloca-tion, the tragedy of a man who can't see where he is going because he doesn't know where he comes from. Ignorant of his true parentage, he flees Corinth when the Delphic oracle foretells that he will kill his father. On his travels, he collides with King Laius at a narrow crossroads and, in a fit of ancient road rage, kills him.

Oedipus goes to Thebes, where he encounters the Sphinx, represented in Greek art as a winged lion with a young woman's head and breasts. She is terrorizing the countryside by devouring young men who can't answer the riddle, "What goes on four legs at dawn, two at noon, three at dusk?" He gives the answer: "Man, who crawls in childhood, walks in maturity, and hobbles with a stick in old age," and the Sphinx throws herself off the cliff below her lair in chagrin. His reward is kingship of Thebes and unwitting marriage to his birth mother, Laius's widow Jocasta. He rules many years, until, as dramatized in Sophocles' clas-sic plays, drought and plague expose his patricide and incest. Then he blinds himself and wanders until the earth finally engulfs him in a sacred grove near Athens.

So, because of its mysterious mythic origins, there are two Sphinxes. The Egyptian Sphinx crouches before the royal tombs, a threat, but also a guardian. If it poses riddles, the answers are meant to maintain order, to keep the lion-sun in its place. The Greek Sphinx ravages the land like a sun out of orbit. Its riddle licenses its destructiveness, and even the riddle's answer licenses its own apparent self-destruction, ironically exposing monarchy to pollution and disaster.

When Chase declared that there is no answer to a California desert Sphinx's riddle, he expressed not only progress-minded civilization's hostility to desert but also its fascinated, half-conscious ambiguity about it. If the desert was a Sphinx, was it the Egyptian one, a guardian of divinely sanctioned nature, or the Greek one, a parvenu trickster threatening progress? Would irrigation make the desert bloom as the rose—a kind of surprise present from the Creator—or would laboriously diverted water merely evaporate and sink into the sand? Or was the desert in some incalculable way both guardian and trickster? Thieves quickly looted the royal tombs that the Egyptian Sphinx supposedly guarded. The Greek Sphinx supposedly gave in to Oedipus by leaping off a cliff—but she had wings.

The twentieth century proved that irrigation can make the California desert bloom, if not as the rose, then as the golf course and other assets. As my initial freeway-view impression of it showed, it began to seem less consequential, less of an obstacle, than it had in the nineteenth. Not even the increasingly noisy creationist movement was interested anymore in the pioneers' notion that desert somehow had a more malign origin than pleasanter biomes. There was less talk of "reclamation" and more of "development," with its associations of "raw land" and "nothing there"—the desert not as wasteland but as nonland.

But then, away from the freeways, an impression of inconsequence could rapidly shift underfoot, as I found in 1983. Now, although mythic ambiguities might seem even more archaic to the twenty-first century's mainstream than they did to the twentieth century's, they have crept back into the discussion of postmodernities like global water shortage and climate change. As we continue to pump California's dwindling aquifers onto the desert's dusty face, we begin to talk less positively of progress and more propitiatingly of "sustainability," a kind of progressive stability or stabilized progress, if such things can be. We wonder what we will do, and what the desert will do, on a drying, warming planet. That brings us back to something like Oedipus's predicament. How can we tell where the desert is going if we don't know where it has been?

In our anxieties, we may be akin to ancient Californians, who also inhabited a drying, warming planet during the past ten to twenty thousand years. We know less about them than we do about ancient Egyptians and Greeks. Still, there is some faint lingering evidence of what they may have thought about the desert.

TWO The Country of Dried Skin

One of the things that can make California desert look inconsequential from freeways is that much of it doesn't even support bushes and lizards. Wide basins of dirty white salts that fill most of the large valleys seem about as close as earthly nature can get to nonentity. Even a big hole in the ground has more character, more promise of life, than desert alkali flats. Their present vacuity makes it hard to credit the fact, proved by the bathtub rings of extinct beaches around them, that they were once fresh-water lakes full of fish and waterfowl and surrounded by pine forests.

In the Mojave's center, archaeologists have found traces of the people who inhabited southeast California toward the end of the lakes' time. Fluted projectile points like those of the Clovis culture that appeared in the Southwest twelve thousand years ago suggest that their makers were hunting sizeable herds of large mammals. They lived around Lake

Mohave, an immense body of water whose dried, salted corpse now stares at the sky along parts of I-15 between Barstow and Las Vegas. The age of the artifacts is unclear, but radiocarbon dating of mussel shells shows that the lake existed ten thousand years ago.

During the next three thousand years, as the lakes shrank and the landscape shifted from grasslands and pine woods to alkaline basins and thorny bushes, those people's descendants must have wondered about the changes. It's hard to imagine what they wondered, but the little we know of their distant successors' complex mythologies perhaps hints at their thoughts.

According to anthropologists Robert Heizer and William Wallace, the origin myths in California's drier areas differed in a fundamental way from those in the humid northwest, which "began with the assumption that the earth was already in existence and looked much as it did in aboriginal times." Since the northwest probably was forested when humans arrived, this suggests that those myths reflected a stable environment. Central and southern California origin myths "reflected a greater interest in the genesis of the world," suggesting that landscapes in those places were more volatile. The latter myths, in fact, have similarities to ones that arose with the spread of Middle Eastern deserts at about the same time.

Origin myths vary from place to place in the California desert, but they fall into two general categories. In the Great Basin and northern Mojave, a divine being created the world from a primeval flood. According to George Laird of the Chemehuevi tribe who lived from the Colorado River to the Tehachapi Mountains, Ocean Woman made the land by rubbing dried skin from her body and sprinkling it overboard while riding a basket boat on the Immortal Water. Two brothers, Mountain Lion and Coyote, helped her to stretch it out and survey its extent, bickering in the process. Mountain Lion, a modest, sensible, straightforward Creator, tried to do things properly but Coyote, a vain, foolish, devious one, kept making a mess of them.

In the southern Mojave and Sonoran deserts, two quarreling Creators also had a hand in making the world, although they emerged not from the Immortal Water but from an atmospheric void. According to

Francisco Patencio of the Cahuillas, a tribe that lived from the San Bernardino Mountains to just west of the Colorado River, brothers who emerged from a floating egg made the world in the course of sibling rivalry. The Tupai-Ipai who lived south of the Cahuillas also told stories of bickering Creator brothers but thought they were the sons of older beings, an earthly mother and a celestial father. Sensible Mountain Lion and foolish Coyote also featured in southern desert myths although as culture heroes rather than Creators.

According to many Indian myths, Coyote was largely if ambiguously responsible for the creation of humans (in some stories, by defecating us), which seems realistic. One might expect him to be blamed for the problematic phenomenon of desert as well. But there seems to be no such myth. In fact, there seems not to be any Native American origin myth that features the desert as such. Many natural phenomena participate in origins, not only water, sky, and animals but celestial bodies, mountains, and plants. But I have seen no origin myth in which "the desert" takes an active part or is even mentioned.

It is not that ancient Californians lacked words for desert. On *The Land of Little Rain*'s first page, Mary Austin declares: "Desert is the name it wears upon the maps, but the Indian's is the better word." She doesn't say what Indians she means or what their word is, but they were probably the Paiutes she knew in Owens Valley. That word may have been like the Chemehuevis', who also spoke a Ute dialect. According to Laird, their word for desert, *tiiravi,* although "a concept of immense importance to Chemehuevi thought," was "never personified or treated as an active noun."

I've seen no explanation of this apparent failure to personify desert in highly anthropomorphic myths. One possibility might be that it was newer than other phenomena. The Immortal Water in many stories hints at the lakes that once existed, some recently. A lake caused by a change in the Colorado's course covered much of the Cahuillas' territory until soon before Europeans arrived. The spreading vacuity of desert might have evoked a more abstract way of thinking than older things. Laird said that *tiiravi* also means the space between one mountain range and another, so it is a unit of measurement as well as a descriptive noun.

Austin's English translation of the "better word"—"the country of lost borders"—also seems abstract.

Maybe a reluctance to personify desert had a darker side as well. Prehistoric people often didn't speak of things they feared, like the dead. Of course, southeast California was their home. "To a white man, the desert is a wasteland," a Chemehuevi told Laird's Anglo wife, Carobeth. "To us it is a supermarket." But markets are tricky things, and most fatal accidents occur in or near the home. Folklore tells of people who, expecting water or food, find only dry springs or empty caches, for which misfortunes, as Austin wrote, there was "no help."

Austin's "country of lost borders" implies anxiety as well as abstraction. The Owens Valley Paiutes feared the "Shoshone Land" to the southeast of their borders. Laird thought that the Chemehuevis, who held territory on the reliable Colorado River, had a more relaxed attitude than "certain other tribes," who found the world a "dark and terrifying place." Anthropologists described the Cahuillas of the drier, hotter desert farther south as having an "all pervasive and intense feeling of apprehension . . . toward the present and the future."

I can relate to that. It is curiously easy to get lost in desert—or simply to feel lost—because it is so open. The way looks clear but distances are deceptive and if the mind wanders during a morning's walk, it can suddenly seem not clear anymore. Then it is noon, and the sun is startlingly hot and there is no shade in sight. I have a recurrent dream of walking through a desert colored garnet, topaz, and beryl like Red Rock Canyon, although it isn't a picturesque badland, just flats, washes, foothills, and peaks. I have a destination but I'm not sure what it is. Sometimes, crossing a ridge, I think I'll reach it on the other side, but then the dream changes. Sometimes I decide to turn back but then I'm unsure of the way.

Mythological speculations are hazy, but I've encountered one material vestige of ancient California that seemed to evoke the Sphinx's ambiguities. Mitchell Caverns State Park in the eastern Mojave does have caves, although I was too absorbed with lizards and bushes on my first visit there to notice. During another trip I took the guided tour. The caves have beautiful limestone formations like nondesert caves except that the underground streams that formed them are dry. They share another feature with caves in more populated places such as Kentucky's Mammoth Cave. They contain a mummy.

As we climbed a metal stairway, the guide pointed to a deep recess and said that the shrouded body of a man left by ancient people lay there. It was the only one in the cave, and the reason for its presence was unknown. It might have been a shaman or other powerful individual placed there so the spirit would continue to preside. The guide speculated further that the man might have inhabited the cave while living, perhaps, like the Delphic oracles, kept there by an anxious society to enhance his prophetic gift.

An early historical visitor to the desert encountered something oddly reminiscent of this. William Manly, a member of a wagon train that wandered into Death Valley in 1849, went hunting one morning and found a cave "which had the appearance of being continuously occupied by Indians." Seeing a "very strange looking track upon the ground," he followed it to "where a small well-like hole had been dug, and in this excavation was a kind of Indian mummy curled up like a dog. He was not dead, for I could see him move as he breathed, but his skin looked very much like the surface of a well-dried venison ham. I should think by his looks he must be two or three hundred years old—indeed he might be Adam's brother and not look any older than he did."

THREE A Cactus Heresy

The first Spanish explorers in California certainly didn't see the desert as a supermarket. Most of them saw as little of it as possible. Pedro Font, a Franciscan priest who accompanied the pioneering Anza expedition to Alta California, wrote vividly of the coast, leaving early descriptions of redwoods, grizzlies, and other wonders. Traversing the Sonoran from the Colorado River to the mountains east of San Diego in December, 1775, he wrote only of scanty or bad water and poor or nonexistent pasturage. The eighteen-day outward crossing made the expedition want to see even less of the desert on their way back. "It was decided to go as directly as possible across the plains and sand dunes which we were following," Font wrote on May 8, 1776, two days after they had begun their return. They reached the Colorado on May 11, leaving behind a trail of starved and exhausted livestock.

There were exceptions to Font's attitude. Another Franciscan, Francisco Garces, made leisurely trips through the Alta California deserts, employing Indian guides and observing their customs with interest. In the spring of 1776, he perhaps was the first European to cross the Mojave, starting from the Colorado River and following native trade routes to the coast. The routes still exist, maintained by ORVs, rolling straight across the *tiiravi* from one sun-baked range to another. Still, Garces was less interested in the desert than in the people he wanted to convert, as Font waspishly observed:

> Father Garces is so well fitted to get along with the Indians and go among them that he appears to be an Indian himself. Like the Indians he is phlegmatic in everything. He sits with them in a circle, or at night around the fire, with his legs crossed, and there he will stay musing two or three hours or more, oblivious to everything else, talking with them with much serenity and deliberation. And although the foods of the Indians are as nasty and dirty as those outlandish people themselves, the Father eats them with great gusto and says they are good for the stomach and very fine. In short, God has created him, as I see it, solely for the purpose of seeking out these unhappy, ignorant, and rustic people.

A member of the more cerebral Jesuit order took a greater interest in the desert itself than the Franciscans, albeit ambivalently. Miguel del Barco, a perceptive observer who manned missions in central Baja from 1738 to 1768, tried to be precise in describing the peninsula. But he couldn't restrain his dismay at its relentless aridity: "Since the land is so elongated it is not strictly speaking uniform in air temperature and soil quality. Be that as it may, one can say in general that its climate is dry and excessively hot, and that the terrain is broken, harsh, and sterile, covered almost completely with stony soil and useless sands, short of rains and natural springs, and thus little suited for livestock, and completely useless for crops and fruit trees which require frequent irrigation."

Despite its barrenness, Barco could not help wondering about the peninsula's diversity of bizarre organisms:

> In the great arroyos of the foothills, and other places, occur various tree species of which several grow straight for two or three varas [yards] and

others up to four or five, and which are less thick than a hand's breadth. But all of these have flimsy, unsubstantial wood. For this reason, and because they produce no fruit that the Indians can use, they are good for little or nothing. The same goes for the little tree called "palo adan" ["Adam tree"—a species of ocotillo] perhaps because it is always naked of leaves and has many long spines. It is true that when rains are good in season, it grows some small leaves; but after a month, they fall, and it spends the year without them. . . . This is a property of almost all the trees and plants in California; they rarely lack spines. The only difference, usually, is that some have more, some less, some larger and some smaller, some curved and some straight.

Barco wondered especially about cactuses, which puzzled him with their ingenious ways of getting and holding water. He described their ribbed structure and succulent pulp accurately but found these characteristics hard to explain. Of the tall, branching "cardon" he wrote:

Even though this tree is permeated with moisture it grows only on dry soil, whether plains or hills. . . . From whence then comes this moisture and fluid of which it is so full? Not from rains, because they are very scanty in California. . . . The cardon nevertheless passes years without rainfall and shows no ill effects; it perseveres with the same serenity, with the same fresh green color and with the same moisture as always. . . . It cannot be said, furthermore, that it makes use of moisture the soil holds at depth. . . . I maintain that this can't be so, because the soil is very dry, and the deeper it is, the drier it gets, so it can't provide the cardon with moisture it lacks. . . . Nor do the cardon's roots go very deep so that it can seek moisture far below the soil surface, because the tree keeps its roots near the surface to the extent that some are visible protruding from the earth. When a hurricane knocks some down, one sees that the roots are near the surface or a little lower. Since the cardon's copious moisture doesn't come from any of these things, then from what could it come? This seems a problem worth the attention of modern savants, so inquiring as skillful investigators of nature. This same problem arises with all the succulent plants.

Another cactus led Barco himself to "inquiring" speculations that ecclesiastical authorities would have found unorthodox, if not heretical. Contemplating the prickly pear, or "nopal," he wondered if such oddities might indicate that the world was not finished as the Bible says it is:

> The author of nature, in producing such a diversity of trees and plants of so many distinct species and properties, gives us to know a little of his infinite wisdom and power. What is more, in producing the nopal, of a nature, properties, shape, and appearance so different from the rest, he opens our eyes to see that . . . he has not exhausted the riches of that wisdom and power, and that there remain new paths to follow, even within natural limits, for newer and newer productions without end, of species not only different from, but contrary to, the rest. Because the nopal, contrary to the nature of other trees, while growing four or five *varas* high, with quite thick branches, contains, even in its trunk, no wood of any kind to support its weight. All the striated types of plants display the extravagant quality of growing with uniform thickness, without coming to a tip at the youngest end, and also of not producing a single leaf in all their lives. The nopal, with renewed extravagance, produces only leaves. From those only are formed its trunk and branches, and the fruit buds off from those same leaves.

Barco was wrong about the nopal's "leaves." The oval prickly pear pads he referred to are really modified stems. He was right in perceiving that cactuses are in the process of diverging substantially from more "normal" plants. Yet his suspicion that organisms might change (albeit under "authorial" control) and that they especially might change in the desert didn't seem to go further than that. Although he saw that desert plant oddities were related to dryness, he didn't suggest that dryness might have caused ordinary plants somehow to develop thorns and succulence. Indeed, he didn't speculate on the cause of Baja California's dryness. He didn't use the word "desert" (*desierto* in Spanish), and in this he was a man of his time. Originally denoting simply uninhabited land, "desert" was not linked to aridity until the nineteenth century.

Barco may have written less than he was thinking as he composed his natural history in Italy from 1780 to 1790. His reference to "modern savants" suggests that he was acquainted with Enlightenment thought, perhaps with the French naturalist Buffon's idea that a bad climate might have caused New World animals like tapirs to degenerate from larger Old World ones like elephants. He may have heard of the ideas associated with another French naturalist, Jean-Baptiste Lamarck, that organisms may change in response to their environment and pass on such changes to their offspring. This is unlikely, since Lamarck didn't publish his theories until some years later. Still, earlier Enlightenment savants such as Charles Darwin's physician-poet grandfather, Erasmus, had similar thoughts.

Reactionary Spain had expelled Barco from Mexico along with his order, however, and the pope had temporarily abolished the Jesuits, partly because of their cerebral bent. Even at liberal Bologna University, the still active Roman Inquisition may have been watching. So the cryptic mention of "inquiring . . . investigators of nature" evidently was as far as he cared to go in writing. The only origins he mentioned were those of the Baja Indians, and he didn't inquire far into those. He recognized their physical similarity to Asians and noted that their myths said they had come from the north after a battle between two great leaders. He didn't record the myths in any detail, concerning himself more as to whether some of their beliefs—in the soul's immortality, for example—might have been acquired from a castaway sailor, although he thought this unlikely.

"For the rest," wrote Barco, "the Indians can make no more sense of their own beliefs than the author has related about them. If one wishes to inquire further, they say that this is what they heard from their elders, and that they know nothing more."

FOUR   The Creator's Dumping Ground

The Anglo explorers who entered southeast California in the mid-nineteenth century had begun to call it desert but otherwise described it much as had their Spanish predecessors. The fur trapper Jedediah Smith was among the first, crossing the Mojave from the Colorado River to Los Angeles in 1827 on roughly the same Indian trade route that Francisco Garces had followed. He called it "a west course fifteen days over a country of complete barrens," where the only water available at times was from "chewing slips of the cabbage pear, a singular plant." Smith found the "barrens" so complete that when the Spanish authorities ordered him to leave California the way he came, he secretly fled north via the San Joaquin Valley instead.

W. H. Emory, a scientifically informed army officer who retraced Anza's route from the Colorado to San Diego in 1846, sounded like Pedro

Font with a few botanical embellishments: "We traveled 'til long after dark and dropped down into a dusthole near two large, green-barked acacias. There was not a sprig of grass or a drop of water and during the whole night the mules kept up a piteous cry for both. There was nothing but the offensive *"Larrea"* [creosote bush] which even mules will not touch when so hungry as to eat with avidity the dry twigs of all other trees and shrubs."

Two weeks later, Emory sounded like Miguel del Barco in a bad mood: "In no part of this vast tract can the rains from heaven be relied upon to any extent for the cultivation of the soil. The earth is destitute of trees, and in great part also of any vegetation whatsoever. A few feeble streams flow in different directions from the great mountains, which in many places traverse this region. These streams are separated, sometimes by plains, and sometimes by mountains, without water, and without vegetation, and may be called deserts, so far as they perform any useful part in the sustenance of animal life."

America's premier explorer, John C. Fremont, had a first impression like mine when he entered the Mojave near Walker Pass in 1843: "Crossing a low Sierra, and descending a hollow where a spring gushed out, we were struck by the sudden appearance of *Yucca* trees, which gave a strange and southern appearance to the country, and suited well the dry and desert region we were approaching. Associated with the idea of barren sands, their stiff and ungraceful forms make them to the traveler the most repulsive tree in the vegetable kingdom."

Fremont was apprehensive, to say the least, at the prospect of crossing the still largely unexplored wasteland, especially when compared to the green coastal landscapes he had just visited: "It was indeed dismal to look upon, and hard to conceive so great a change in so short a distance. One might travel the world over without finding a valley more fresh and verdant—more floral and sylvan—more alive with birds and animals—more bounteously watered—than we had left in the San Joaquin: Here within a few miles, a vast desert plain spread before us, from which the boldest traveler turned away in despair."

Coming from the lush eastern United States, instead of semiarid Spain like Font and Barco, Fremont found the "vast desert plain" not just frightening, but startling, bizarre. "The whole idea of such a desert . . . is a novelty in our country, and excites Asiatic, not American, ideas. Interior basins with their own systems of lakes and rivers, and often sterile, are common enough in Asia . . . but in America such things are new and strange, unknown and unexpected, and discredited where related."

Fremont got his fill of the novelty in the next month: "Travelers through countries affording water and timber can have no conception of our intolerable thirst while journeying over the hot yellow sands of this elevated country, where the heated air seems to be eternally deprived of moisture." It was a dismal discovery for a seeker after profitable resources. The notion of the desert as a supermarket did not appeal to him, as he made clear in describing a band of what he called "desert Arabs" who stalked his party of fur trappers and frontiersmen: "Many of these Indians had long sticks, hooked at the end, which they used in hauling out lizards, and other small animals, from their holes. During the day, they occasionally roasted and ate lizards at our fires. . . . On the following day, we . . . continued on our way through the same desolate and revolting country, where lizards were the only animals, and the tracks of lizard eaters the principal sign of human beings."

Along with revulsion, however, Fremont had an uneasy sense akin to Barco's that puzzling and marvelous things occurred in the strange landscape:

> The country had now assumed the character of an elevated and mountainous desert; its general features being black, rocky ridges, bald, and destitute of timber, with sandy basins between. . . . But, throughout this nakedness of sand and gravel, were many beautiful plants and flowering shrubs, which occurred in many new species, and with greater variety than we had been accustomed to see in the most luxuriant prairie countries; this was a peculiarity of the desert. Even where no grass would take root, the naked sand would bloom with some rare and rich flower, which found its appropriate home in the arid and barren spot.

In one place, "twenty miles to the southward, red stripes of flowers were visible during the morning, which we supposed were variegated sandstones. We rode rapidly during the day, and in the afternoon emerged from the yucca forest at the foot of an outlier of the Sierra before us, and came among the fields of flowers we had seen in the morning, which consisted principally of the rich orange-colored California poppy, mingled with other flowers of lighter tints."

A man of action, Fremont did not ponder the reasons for the desert's "peculiarity," although he brought back some of the first scientific specimens of the shrubs whose diversity impressed him. John Torrey, a botanist, used the specimens to describe and name the new species. Fremont also collected fossils of Mesozoic ferns, but this hint that the dry West might once have been wetter did not prompt him to wonder in print how the dryness and its peculiar organisms arose.

Six years after Fremont crossed the Mojave, William Manly's wagon train became the archetype of California desert revulsion and incomprehension when it got stuck in Death Valley. Manly, a Vermont farm boy and Wisconsin backwoodsman, described a toxic landscape wherein days passed without sight of a living animal: "As I reached the lower part of the valley I walked over what seemed to be boulders of various sizes, and as I stepped from one to another, the tops were covered with dirt, and they grew larger as I went along. I could see behind them, and they looked clear, like ice, but on closer inspection proved to be immense blocks of rock salt, while the water that stood at their bases was the strongest brine."

Trapped in stranger places than even Fremont had seen, Manly and his party felt compelled to speculate on desert origins. As with Barco's cactus heresy, their ideas might have shocked orthodox religionists back in civilization:

> One fellow said he knew this was the Creator's dumping ground, where he had left the worthless dregs after making the world, and the devil had scraped these together a little. Another said this must be the very place where Lot's wife was turned into a pillar of salt, and the pillar [had] been broken up and spread around the country. He said if a man was to die he would never decay, on account of the salt. Thus the talk went on, and it seemed as if there were not bad words enough in the language to properly express their contempt and bad opinion of such a country as this.

Manly's own Death Valley meditations were less fanciful but equally heterodox:

> If the waves of the sea could flow in and cover its barren nakedness . . .
> its exhausting phantoms, its salty columns, bitter lakes, and wild, dreary,
> sunken desolation . . . it would be indeed a blessing, for in it there is
> naught of good, comfort, or satisfaction, but ever in the minds of those
> who braved its heat and sands, a thought of a horrid charnel house, a
> corner of the earth so dreary that it requires an exercise of strongest
> faith to believe that the great Creator ever smiled upon it as a portion
> of his work and pronounced it "very good."

Manly went into the desert a levelheaded man, as his eventual success in getting most of his starving party out demonstrated. He emerged into coastal California's Arcadian valleys sounding more than a little addled:

> There before us was a beautiful meadow of a thousand acres, green
> as a thick carpet of grass could make it, and shaded with oaks, wide-
> branching and symmetrical, equal to those of old English parks, while
> all over the low mountains that bordered it to the south and over the
> broad acres of luxuriant grass was a herd of cattle numbering many
> hundreds, if not thousands. . . . All seemed happy and content, and
> such a scene of abundance and rich plenty and comfort bursting thus
> upon our eyes, which for months had seen only the desolation and
> sadness of the desert, was like getting a glimpse of paradise, and tears
> of joy ran down our faces.

FIVE

An Evolutionary Backwater

As I've said, the California desert has connections with others. One of the vivider landscape surprises I've experienced was of starting a trek in the Peruvian Andes and finding that the trailhead vegetation—on a bench above the Urubamba River at nine thousand feet—might have been in Red Rock Canyon or the Providence Mountains. The plants weren't the same but they were similar variations on the themes of spikiness, hairiness, smelliness, and so on. I saw what looked like ephedra, cholla, cereus cactus, and prickly pear. The only marked difference from the Alta Californian Sonoran was the presence of slender tree cactuses, like toy saguaros.

I shouldn't have been surprised. South America has high mountains and midlatitudes, so it has many deserts. Their ubiquity prompted one young observer to entertain some of the first thoughts about natural as opposed to divine or devilish desert origins. Not surprisingly, that observer was Charles Darwin, who saw plenty of cactus and creosote bush while crisscrossing Argentina, Chile, and Peru from 1833 to 1835. Like Miguel del Barco and John Fremont, he found the experience grim but strangely suggestive.

Darwin was ecstatic about South America's rain forest, but troubled by its desert. "I am tired of repeating the epithets barren and sterile," he wrote in the journal later published as *The Voyage of the Beagle*. "While traveling through these deserts one feels like a prisoner shut up in a gloomy court, who longs to see something green and smell a moist atmosphere." When he came upon a stand of fossil trees similar to the living ones of damp Tierra del Fuego in an "utterly irreclaimable and desert" Chilean valley, he wondered at the "vast, and scarcely comprehensible" changes implied by the petrified trunks, changes that he thought must have occurred since the rise of the Andes:

> It has been inferred with much probability that the presence of woodland is generally determined by the amount of moisture. . . . Confining our view to South America, we should certainly be tempted to believe that trees flourished only under a humid climate; for the limit of forestland follows, in a most remarkable manner, that of the damp winds. In the southern part of the continent, where the western gales charged with moisture from the Pacific prevail, every island of the broken western coast . . . is densely covered with impenetrable forests. On the eastern side of the cordillera, over the same extent of latitude, where blue sky and a fine climate prove that the atmosphere has been deprived of moisture by passing over the mountains, the arid plains of Patagonia support a most scanty vegetation.

Darwin's petrified stumps raised a basic problem of early nineteenth-century science. Growing evidence, rock strata and fossils, showed that the earth had undergone vast changes. Rocks were not necessarily at odds with religious orthodoxy, which encompassed floods and earth movements. But fossils different from living organisms had troubling implications for the biblical doctrine that life is a divinely finished creation. Although he inferred that mountains and climate caused aridity, not divine neglect or devilish mischief, Darwin was an orthodox

Anglican during the *Beagle* voyage and resisted the further inference that dryness had somehow caused the spiky shrubs growing around his petrified trees.

Darwin's *Beagle* journal had little to say about desert's possible effects on life, or about desert life in general. He did note in northern Chile that only one animal, a snail, seemed abundant on a particularly barren plain: "In the spring a humble little plant sends out a few leaves, and on these the snails feed. . . . I have observed in other places that extremely dry and sterile districts, where the soil is calcareous, are extraordinarily favorable to land shells." He didn't speculate on the reasons for this snail abundance, instead passing on to the premechanized desert travelers' obsession with "corn and straw for our horses."

Darwin expressed puzzlement at finding that superabundant reptiles—tortoises and large lizards—were the main herbivores on the "very sterile" Galapagos Islands:

> We must admit that there is no other quarter of the world where this Order replaces the herbivorous mammalia in so extraordinary a manner. The geologist on hearing this will probably refer back in his mind to the Secondary epochs, when lizards, some herbivorous and some carnivorous, and of dimensions comparable only to our existing whales, swarmed on the land and in the sea. It is, therefore, worthy of his observation, that this archipelago, instead of possessing a humid climate and rank vegetation [like the "Secondary epochs"], cannot be considered otherwise than extremely arid, and, for an equatorial region, remarkably temperate.

Again, however, he passed over the oddity, jumping in a paragraph from "ugly . . . stupid . . . lazy" land iguanas that fought over cactus branches "like so many hungry dogs with a bone" to "fifteen kinds of sea-fish."

Darwin continued to shy away from desert after he abandoned creationism. In *The Origin of Species,* he interpreted desert, in passing, as an evolutionary backwater where aridity has muted the driving force of organic change—natural selection of favorable variations through competition. To be sure, competition would exist among organisms able to survive in a desert—the Galapagos land iguanas had been competing for cactus branches. Natural selection would cause an animal or plant that needed less water than its rivals to produce more offspring. That offspring, presumably, would resemble it in needing less water. But the more arid a place was—the less life it could support—the more the "struggle for life" would be "almost exclusively with the elements."

Struggle with the elements was not strictly competition for Darwin: "Two canine animals, in a time of dearth, may be truly said to struggle with each other which shall get food and live. But a plant at the edge of a desert is said to struggle for life against the drought, though more properly it should be said to be dependent on the moisture." Adaptation to a harsh environment seemed to provide little scope for natural selection compared to the species-packed rain forests and coral reefs Darwin preferred, and he didn't discuss particular desert plants or animals in *Origin*, a book otherwise remarkable for its exhaustive detail.

In *The Descent of Man*, Darwin used desert animals as a negative example. He considered sexual selection a major form of competition, with brilliant male bird plumage a case in point, except in the case of desert: "In regard to birds that live on the ground every one admits that they are coloured so as to imitate the surrounding surface. . . . Animals inhabiting deserts offer the most striking cases for the bare surface affords no concealment, and nearly all the smaller quadrupeds, reptiles, and birds depend for safety on their colours. . . . Calling to my recollection the desert-birds of South America . . . it appeared to me that both sexes in such cases are generally coloured nearly alike."

Darwin might have concluded from his ideas that, since deserts are evolutionary backwaters, the organisms in them must be very old. When writing *The Origin of Species* he assumed that millions, even billions of years would have been required for natural selection to produce life's astonishing diversity. But this would have conflicted somewhat with his South American observations, which suggested that a fairly recent and rapid elevation of the Andes was causing deserts there. Darwin saw that the earth had actually risen after Chilean earthquakes, and he described finding ancient buildings in a valley now too dry for human habitation. This implied that plants and animals would have had to adapt quickly to the increasing dryness or disappear like the ancient buildings' human inhabitants. That he didn't try to resolve this apparent paradox is a measure of his lack of interest in desert.

Lack of interest in desert organisms typified nineteenth-century Darwinism. Darwin's friends and supporters, like Thomas Henry Huxley and Joseph Hooker, also gravitated to lush places—northern Australia, New Guinea, New Zealand, the Himalayas—and saw competition between organisms as the main force of evolution. Natural selection's co-discoverer, Alfred Russel Wallace, spent more than a decade in South America's and Asia's rain forests, not a day in their deserts. When he cited desert as a habitat in his exhaustive *Geographical Distribution of Animals*, Wallace lumped all the North American ones into a "Central or Rocky Mountain Sub-region" that extended from Midwest prairies to the Sierra Nevada and south to central Mexico. This echoed a nineteenth-century habit of calling all land west of the Mississippi "the great American desert" without regard to the enormous differences among the West's grasslands, scrublands, semiarid woodlands, and real deserts.

Wallace's brother, who had immigrated to California, sent him a horned lizard that survived the trip and lived for a few months in England. But even that didn't seem to pique the great naturalist's interest in desert life. When he visited his brother in the 1880s, he rushed to see redwood forests and mountain meadows, not cactuses and creosote bushes.

SIX Anti-Darwinian Lacertilians

One reason why Darwin found the Galapagos Islands "very sterile" was that most of their terrain is recently erupted lava: "Nothing could be less inviting than the first appearance. A broken field of black basaltic lava, thrown into the most rugged waves, and crossed by great fissures, is every where covered by stunted, sun-burnt brushwood, which shows little signs of life." He agreed with the *Beagle*'s captain, Robert FitzRoy, that the lava gave the islands an infernal aspect, making their swarms of blackish marine iguanas seem like imps of darkness on the shores of Pandemonium despite the big lizards' inoffensive diet of seaweed.

Lava flows are a feature of California's deserts, and I can understand Darwin's impression. Whether thrusting out of the ocean or smeared across continental mesas, they can indeed seem hellish. A closer look at Cinder Cone Lava Beds in the Mojave National Preserve gave me a somewhat different impression than Darwin's, however. The beds looked infernal enough from a distance, but when I got into their complex of box canyons the landscape began to seem almost cozy.

Floods and wind had abraded the lava to a smoothness that felt silky in the fresh April air. Sun-warmed black chunks protruding from the white sand of the washes made comfortable recliners. The cubical basalt ledges above were like the remains of cyclopean walls, stairways, corridors, and antechambers, but without the disturbing grotesqueries of Red Rock Canyon's hoodoos. Although the lava blocks seemed good habitat for the fat black chuckwallas that might have mimicked Darwin's marine iguana imps, I didn't find the big lizards there.

The lizards I did find didn't seem ugly, lazy, or stupid, but graceful, busy and rather astute. Passing a fragrant shrub with purple flowers full of bees, I saw a pale shape zip out of its shade and then apparently disappear. I kept watching, and eventually a zebra-tailed lizard that had blended with the sand when it stopped running decided to move again. It raised its tail and sped toward a cheesebush, where it began swelling out its throat and doing push-ups that flashed a bright blue patch marked with two black bars on its flank.

Soon, another smaller zebra-tailed lizard seemed to get the message, since it emerged from a nearby shrub and dashed away. Then another, slightly larger one emerged from the cheesebush and started wriggling the tip of its tail as though to get the first lizard's attention. Clearly attracted by this "come-hither" gesture, the first lizard obligingly approached and began swelling its throat and doing push-ups again, watched intently by the newcomer. Before very long, they ran into the cheesebush together.

The trio clearly had exchanged a lot of information. April is a benign time of year, of course, but even so the lizards seemed much more concerned with their effects on each other than with the desert's effects on them. Competition evidently was underway, although whether for space or sex, or both, was unclear. Indeed, the competition was more conspicuous than it would have been in woods or grassland. Darwin's evolutionary backwater assumptions about desert didn't seem upheld.

Those assumptions weren't universally upheld even in the nineteenth century. When Darwin's friend and supporter, the Harvard botanist Asa Gray, published a survey of North American vegetation in 1884, he followed Darwin's lead in attributing a recent rise to the Americas' western mountains. But he assigned an active evolutionary role to desert life. He wrote that "the peculiar elements of the California flora, and also of the southern Rocky Mountain region and the Great Basin" probably had originated in "the Mexican Plateau." That flora had then spread north and—because of growing aridity as the West Coast mountains rose within the past few million years—replaced what he called an "Arcto-Tertiary" flora of temperate conifers and hardwoods that had covered much of the Northern Hemisphere since the dinosaur age.

Gray was ambivalent about natural selection's role in this northward spread of desert. He thought that "competition from the Mexican Plateau vegetation" had contributed to the Arcto-Tertiary forest's demise in the Southwest, but added: "It is unnecessary to build much upon the possibly fallacious idea of increased strength gained by competition. Opportunity may account for more than exceptional vigor." Without the opportunity of drying climate, in other words, cactuses and creosote bushes wouldn't have been much competition for aboriginal conifers and hardwoods.

Gray's ambivalence about competition reflected increasing doubts about Darwinism toward the nineteenth century's end. Natural selection seemed too random and cumbersome to have produced living organisms' exquisite adaptations, especially after 1866, when the leading English physicist William Thomson, later Lord Kelvin, "proved," from the rate at which the earth's molten center was thought to be cooling, that the planet could not be much more than a hundred million years old. Given such a restricted time frame, many naturalists looked with renewed interest on the idea of Enlightenment savants like Lamarck that environment might have direct effects in changing organisms.

The Reverend George Henslow, a botanist and nephew of one of Darwin's university professors, used desert as a case in point for a more direct and prompt cause of biological change. Noting Darwin's reticence about desert, Henslow wrote: "He bases his theory of Natural Selection on *the struggle between living organisms,* and to a much less degree on the environment. My contention is that precisely the reverse really obtains; that morphological adaptations arise, and are maintained—not in consequence of any struggle with other living beings of the same or of different species—but solely through the direct action of the new environment."

Henslow thought desert plants' thorniness, leaflessness, and hairiness demonstrated those features' environmentally induced origins: "We thus begin to suspect, indeed very strongly, that the various peculiarities (such e.g., as the densely hirsute clothing and the consolidation of the vegetative tissues) are the direct result of the dry climatic conditions surrounding the plants, and that the unfavorable environment actually brings about the production of just those kinds of structures which are best able to resist the injurious effects of climate, and so enables the plant to survive under them."

Another reason scientists backed away from Darwin was that he couldn't supply a credible mechanism whereby organisms might produce and transmit the variation upon which natural selection could work. As one anti-Darwinian, the American naturalist Edward D. Cope, complained, natural selection alone might explain the "survival of the fittest," but it didn't explain the "origin of the fittest." It might explain why cactus replaced broadleaf trees in desert; it didn't explain how cactus developed succulence, spines, and other variations useful in dry climate.

If desert plants became succulent or spiny simply because the heat and dryness affected them that way, as Reverend Henslow maintained, then that supplied at least part of the mechanism. But Henslow's idea had a problem. Cactuses grown in moist gardens or hothouses reproduce little cactuses as spiny and succulent as their parents, so these traits, even if dryness originated them, somehow have become inheritable. Many nineteenth-century scientists, including Cope, also adopted the Lamarckian idea that acquired traits might be inherited, thus completing a mechanism for variation.

Cope conceived elaborate "neo-Lamarckian" theories about the acquisition and inheritance of variations, although, being a zoologist, he did not apply them to desert bushes. In fact, he was the "founder of American herpetology," the first in the United States to study reptiles and amphibians systematically. His live-in lab contained a cold-blooded menagerie, including a Gila monster that apparently liked to have its head scratched. Cope frequently collected in the Southwest, and he described many species. He was so adept at this that he once classified a newly discovered lizard by simply examining it in another naturalist's hand. He was a paleontologist too. He described the first North American fossil "lacertilian" from an early Age of Mammals deposit in the West in 1867. Another paleontologist, Joseph Leidy, had described lacertilian fossils earlier, but they'd turned out to be dinosaur bones once the distinction between lizards and dinosaurs (unknown to Darwin when he visited the Galapagos in 1835) became evident.

Cope might have used reptile herbivores like those in the arid Galapagos to argue against Darwin's tacit assumption that evolution would be slow in desert. According to his anti-Darwinian "origin of the fittest" theories, a small omnivorous animal like a lizard might become a larger herbivorous one more quickly than natural selection alone would allow. Finding itself in a place like the Galapagos without large mammal herbivores, it would not simply continue its modest omnivory, but would begin to eat more plants, a process Cope called "archaesthicism." This change of diet would affect its anatomy—jaws, teeth, digestive system, even size—a process Cope called "kinetogenesis." Then it would bequeath these new traits to its offspring through "recapitulation," whereby changes in the parent's body would affect the embryo.

Cope might have pointed out, furthermore, that Darwin's belief that the volcanic Galapagos rose from the Pacific fairly recently implied that the islands' unique reptiles had evolved their unusual traits not slowly, but quickly. (Darwinism's leading American foe, Harvard biologist Louis Agassiz, visited the Galapagos in 1872 and cited their recentness and unique biota as evidence against gradual natural selection.) Cope then could have posited a similar evolution for the North American desert's plant-eating lizards. Most large mammal herbivores might have died out as scattered, spiky bushes replaced Asa Gray's lush Arcto-Tertiary vegetation, providing an opportunity for reptiles to eat more plants and grow larger. Six-inch insect-eating lizards could have become eighteen-inch chuckwallas in a few hundred thousand years.

Scientific competition, however, came between Cope and thoughts on desert evolution. His 1867 lizard fossil was so fragmentary that it was assigned to a better-preserved genus that another paleontologist, Othniel C. Marsh, found and named in 1871. A wealthy Yale University professor, Marsh could exploit western fossil deposits with large, well-funded expeditions, while Cope, usually self-employed, had to explore alone or with a few underpaid assistants. Cope envied Marsh's prominence; Marsh envied Cope's brilliance. Their rivalry grew so intense that it descended to ad hominem newspaper attacks and spying on each other's collections and fossil sites.

Marsh's 1871 discovery did briefly spur the rivals' fossil lizard collecting, as Charles Gilmore, a Smithsonian herpetologist observed: "Owing to the continued exploration of the Eocene of the Bridger Basin in southeastern Wyoming and of the Oligocene deposits in northeastern Colorado by Cope and Marsh, knowledge of the extinct lizards increased more rapidly in the two years following, 1872 and 1873, than in any other similar period in the entire history of the country. No less than 10 genera and 26 species were established by Leidy, Cope and Marsh, within this period."

But neither Cope nor Marsh did more than describe, name, and classify their fossil lizards. They were much too busy using more impressive fossils to promote their clashing agendas. Cope applied his ideas about omnivores quickly evolving into herbivores not to lizards but to mammals. He thought they had been largely arboreal when the dinosaurs ruled (he and Marsh were the first to discern the basic outlines of dinosaur evolution), and then had descended from the trees and evolved into large ungulate herbivores after the dinosaurs' extinction. Marsh was a confidant of Darwin and T.H. Huxley, who furthered their agenda by finding and publicizing important "missing links" between groups of organisms. He used his huge fossil collection to develop a Darwinian lineage of horse evolution that made him famous, to Cope's chagrin, and is still used in biology classes.

Despite their differences, both paleontologists' primary criterion for measuring evolutionary development was the anthropomorphic one of brain growth. Cope thought mammal brains had grown through inheritance of traits acquired by interaction with the environment. Marsh thought they had grown through natural selection of larger-brained individuals. Neither man applied such progressive notions to reptiles that Darwin had dismissed as stupid. Cope read and admired *The Voyage of the Beagle* at an early age, and its disdain for desert perhaps influenced him. A comment he made about crossing the Great Basin Desert on a train was typical: "It was very hot and dusty and I do not wish to visit such a country in the summer. One has alkali, sage brush and grease wood and vice-versa, all the way." His trips to the Southwest often continued into Mexico's tropical forests, which he described almost as ecstatically as Darwin had the Amazon.

Cope published a 1,117-page tome on North American reptiles that noted chuckwalla and desert iguana plant eating but failed to consider its origin. Marsh was interested in lizards mainly as a way to annoy Cope. In 1872, when their rivalry reached one of its periodic peaks (they would squabble until both died in the 1890s), the Yale professor described sixteen new species in five genera of fossil lizards he'd found on his 1871 expedition. One of his genera, *Oreosaurus*, included several species that Cope placed in a genus of his own, *Xestops*. Cope's pride in his classifying skills made such coining of rival names a sure way for Marsh to torment him.

Marsh's sixteen new species manifested another reason beside backward brains for the rivals' lack of interest in lizard evolution. Whereas many of their early mammals could be traced through fossil links to familiar living beasts like horses, only one of the fossil lizards they found seemed to have a living relative in North America, the Gila monster of the genus *Heloderma*. The rest belonged to groups of uncertain identity. A decade after Marsh's annoying *Oreosaurus* publication, this uncertainty allowed Cope to have the disparaging last word on his rival's fossil lizard species: "As Professor Marsh does not give us any clue as to the affinities of these forms, they cannot be considered further."

The two paleontologists' bad attitudes seem to have set the tone for early lizard evolution studies. Charles Gilmore noted that not even Cope had set out deliberately to find fossil lizards: "All of the fossil lizard remains known from North America, if the Mosasauria [giant marine lizards] are excluded, have been discovered as an incidental part of the work of various expeditions organized either for geological exploration or for the collecting of other classes of fossil vertebrates. Nowhere is there a record of an expedition having as its primary object the securing of lizard specimens." For decades after Cope and Marsh's 1870s discoveries, Gilmore lamented, scientists had not described a single new fossil lizard.

SEVEN

Descriptive Confusion

Edward Cope and Othniel Marsh collected in coastal California but not in its desert, perhaps thinking it too barren even for bone hunting. Early scientists who did explore it were ambivalent about its evolutionary role. Some saw it as a backwater; some saw it as filled with competition. Most felt they had enough to do describing the wretched place, much less trying to guess at its past. Petrified wood, shells, and other evidence lay about, unhidden by vegetation, but desert fossils can be confusing. When I was at Mitchell Caverns, many black, scatlike cylinders on rocks suggested that small, living carnivores were inordinately common until I went to the visitor center and saw the cylinders on Paleozoic limestone in a display case. They were fossil corals.

Pioneer geologists like Clarence King, who worked for Josiah Whitney's California survey in the 1860s, agreed with Darwin and Asa Gray that

fast-rising mountains had changed the climate by cutting off ocean moisture and that desert had replaced ancient forests and waters within the
past few million years. Traveling on a mule from the Colorado River to
Los Angeles in 1866, King thought the Salton Sink's "level floor, as white
as marble," was the bed of "an ancient sea—a white stain defining its
western margin as clearly as if the water was just receded."

But life's history seemed almost irrelevant to the vast cataclysms of
landscape formation that King envisioned: "Sheets of lava poured down
the slopes of the Sierra covering an immense proportion of its surface,
only the high granite and metamorphic peaks reaching above the del-
uge. Rivers and lakes floated up in a cloud of steam and were gone
forever. The misty sky of these volcanic days glowed in innumerable
lurid reflections, and, at intervals along the crest of the ridge, great cones
arose, blackening the sky with the plumes of mineral smoke."

Looking back from San Gorgonio Pass as he escaped into the "gentle
golden green" coastal landscape, King saw California's Sonoran Desert
as an extinct world's mummified remains. He described it in terms
reminiscent of gruesome Civil War battlefield photographs he'd recently
seen back east: "Spread out below us lay the desert, stark and glaring,
its rigid hill chains lying in disordered groupings, the attitudes of the
dead. The bare hills were cut out with sharp gorges, and over their
stone skeletons scanty earth clings in folds, like shrunken flesh; they are
emaciated corpses of once noble ranges now lifeless, outstretched as in
a long sleep. Ghastly colors define them from the ashen plain in which
their feet are buried."

Pioneer biologists' descriptions were less grandiose, but most dwelled
little more on desert life's past. William Brewer, a botanist with Whitney's
survey, wondered at desert bushes' ingenious durability:

> The California deserts are clothed in vegetation—peculiar shrubs
> which grow one to five feet high, belonging to several genera, but
> known under the common names of "sagebrush" and "greasewood."
> They have but little foliage, and that of yellowish gray; the wood is
> brittle, thorny, and so destitute of sap that it burns as readily as other
> wood does when dry. Every few years there is a wet winter, when the
> land of even these deserts gets soaked. Then these bushes grow. When

it dries, they cease to put forth much fresh foliage or add much new wood, but they do not die—their vitality seems suspended. A drought of several years may elapse, and when at last the rains come, they revive into life again!

Despite his exclamation point, however, Brewer showed no curiosity as to how the bushes had gotten that way.

Encountering the Great Basin Desert east of Yosemite in 1869, John Muir wondered at the arid landscape's floral diversity: "I found the so-called desert of Mono blooming in a high state of natural cultivation with the wild rose, cherry, aster, and the delicate abronia [sand verbena]; also innumerable gilias, phloxes, poppies, and bush compositae. I observed their gestures and the various expressions of their corollas, inquiring how they could be so fresh and beautiful out in the volcanic desert. They told as happy a life as any plant company I have ever met, and seemed to enjoy even the hot sand and wind." He didn't wonder enough to inquire further, however, and turned back into the mountains.

Decades later, the first substantial government-sponsored attempt to study California desert biology, the 1891 Death Valley Expedition, continued to favor description. Frederick V. Coville, in his report on botany, repeated John Fremont's observation that desert plants spilled over the passes leading from the San Joaquin Valley into the Mojave: "It is here that the desert vegetation changes to the chaparral of interior California. Several characteristic desert species were found to occur on the west slope of Walker Pass, some of them as low an altitude as 2,800 feet." Along with Joshua trees, Coville listed fourteen desert bush species he'd observed west of the pass.

"It is in the shrubby vegetation of the desert that the arid character of the region is most conspicuously reflected," he concluded, citing an ominous example of that character: "the excessive dryness of the region was evidenced by the fact that the pencil marks on a roadside grave board, which had been twelve years exposed to atmospheric effects, still appeared clear and fresh with the wood retaining its natural appearance, not changing to the gray color of weathered timber."

The most prolific early desert describer was Willis Linn Jepson,

California's leading field botanist until his death in 1946. Born on a Vacaville ranch in 1867, a product of the agrarian arcadia that pioneers like William Manly established, Jepson got the first University of California, Berkeley, botany PhD in 1898. He often visited the desert, collected most of its plants, and described their adaptations concisely in his 1,238-page *Manual of the Flowering Plants of California:*

> The vegetation has the characteristic aspect or facies of plants of desert regions, that is, there is everywhere exhibited a marked development of structures to inhibit transpiration, or physiological devices for the conservation of water. These various forms may be described, in general, under five headings: (a) Plants with condensed bodies such as the species of *Cereus, Mammilaria, Echinocactus,* and *Agave.* (b) Plants with reduced or obsolete [vestigial] leaf surface, such as *Cercidium torryeanum* [paloverde], *Ephedra californica,* and *Parosela spinosa* [smoke tree]. (c) Plants with fleshy leaves, such as *Lycium andersonii.* (d) Plants with resinous, wooly, or scurfy covering to the whole body, such as *Atriplex hymenelytra* [desert holly], *Grayia spinosa* [hopsage] and *Tidestromia oblongifolia* [Arizona honeysweet]. (e) Ephemeral annuals. . . . Certain shrubs or small trees show a similar adaptation in that their leaves appear only during the rains (*Fouquieria splendens* [ocotillo]).

Jepson didn't wonder how the adaptations had evolved, however. He agreed with Asa Gray that desert plants had "their relationships and origins mainly within the regions southward, chiefly in northern Mexico," and that they, along with California's other vegetation, probably started to reach their present forms during the Pliocene epoch, when there was "an elevation of the whole west coast." Otherwise, his voluminous publications and field notes show no interest in how desert organisms acquired their "various forms." He reflected that he had an "irresistible fascination" but no "native affection" for the desert because of its "ruthless, implacable defiance." Echoing Muir, he concluded that a diverted Colorado River would "furnish the breeding spot for another great civilization to match the Nile or the Mesopotamia of old."

Some thoughts about origins had begun to arise. Frederick Coville, the Death Valley Expedition botanist, echoed Darwin in finding less competition in desert than in other habitats. He saw this as a scientific advantage:

The scantiness of the desert vegetation possesses more significance than has ever been attributed to it. Except in rare instances the individual shrub is separated from its nearest neighbor by a distance of several meters. Never do they stand so close together as to crowd or shade each other. . . . In a humid climate, the individual plants, whether trees, shrubs, or herbs, grow close together and by reason of the shade thus produced tend to deprive each other of light and of heat. The result is an active competition of both individuals and species for these necessaries of life. Between the separated individuals of the desert, however, no shade is produced, and they carry on, therefore, no struggle against each other for heat and light.

It is evident therefore that desert shrubs essentially present to their environment the anomaly of a struggle for existence, not against other plants, but against nonorganic physical forces alone. This fact makes the study of their adaptations especially interesting and instructive, for one element in the usual complexity of environment is removed, and we are able to perceive the simple influence of climatic and soil conditions.

On the other hand, the report compiled by the expedition's leading zoologist, C. Hart Merriam, recorded plenty of competition among desert animals. Concerning the zebra-tailed lizard, then called the "gridiron-tailed" lizard, the report inferred that it was "very much more abundant than any other species" because it was the hardest to catch. "It starts off at full speed," Merriam wrote, "as if fired from a cannon and stops with equal suddenness, thus escaping or eluding its enemies; the coyotes, hawks, and larger lizards. When running it moves so swiftly that the eye has difficulty in following, and when at rest its colors harmonize so well with those of the desert that it can hardly be seen."

The zoologists were equally impressed with "the ferocity and greed" of the leopard lizard. They found one individual with a full-grown spiny lizard two-thirds its size in its stomach, another with a full-grown horned lizard and the remains of another leopard lizard.

Yet this obvious "struggle for existence" did not provoke thoughts as to how dryness might have affected lizard speed or greed, not in the expedition report, anyway. The only expedition zoologist to speculate about causes was Robert E. C. Stearns, a curator of mollusks at the National Museum who, like Darwin, was impressed by the ubiquity of snails in limestone desert. "*Bythinella protea* is an exceedingly variable form," he wrote of a species that inhabits desert springs, "including examples that have a perfectly smooth surface, and others that are variously sculptured." Stearns thought that dissolved chemicals in the spring water caused the variation in the snails' shells, with smooth shells growing in dilute solutions, and sculptured shells growing in water "supersaturated with the bitter chemicals."

He saw a "struggle for existence" in this, because snails with sculptured shells would have demonstrated superior fitness to inhabit very salty or alkaline water. As Darwin had thought, the struggle would be more with the harsh desert environment than with other snails. At the same time, however, Stearns's idea that spring water's mineral content caused the variation in their shells led away from Darwin's evolution through natural selection of inherited traits toward Reverend Henslow's and Coville's ideas of evolution through direct action of the environment. Like a traveler without a compass on a vast scrubby plain, early California desert biology seemed to be going in circles.

EIGHT

A Murderous Brood

For all its extremes, the California desert can be surprisingly benign at times. The first night I spent at Mitchell Caverns was a complete change from the chill, windswept grotesqueries of Red Rock Canyon the night before. The evening was warm and calm, rare in spring. The creosote flats rose toward the dove gray Providence Mountains with a kind of tender majesty beneath a royal blue sky and a full moon that was as inviting in its gigantism as the previous night's had been daunting. Strolling over the white sand was like floating. The only disturbers of the peace were crickets and faint poorwill and great horned owl calls from the canyons.

Such charms tempted nature lovers into the desert as well as explorers and scientists when access grew easier. Some began to speculate more freely than the scientists about the desert's ancient past. While biologists backed away from natural selection because of its vagueness about things like variation, Darwinism's ideological adherents like the sociologist Herbert Spencer took advantage of the vagueness to extend natural selection to nonbiological phenomena. Spencer saw it as the principle underlying human culture as well as organic evolution, with "the survival of the fittest" leading toward civilized progress. A mélange of Darwinism and Spencer's "Social Darwinism" shaped public attitudes toward the desert as intellectuals like Mary Austin spilled eastward over the passes and ran headlong into the "tormented, thin forests" of Joshua trees.

Austin was one of the first writers to like desert, and her enthusiasm led her toward a sensuous mysticism, toward feeling "a palpable sense of mystery in the desert air." In a book published in 1909, *Lost Borders*, she wrote, "If the desert were a woman, I know well what like she would be: deep-breasted, broad in the hips, tawny, with tawny hair, great masses of it lying smooth along her perfect curves, full lipped like a sphinx, but not heavy-lidded like one, eyes sane and steady as the polished jewel of her skies . . . and you could not move her, no, not if you had all the earth to give, so much as one tawny hair's breadth beyond her own desires."

But, although living semi-isolated in desert towns, Austin was well aware of early science. Frederick Coville's botanical descriptions influenced *The Land of Little Rain:* "It is recorded in the report of the Death Valley expedition that after a year of abundant rains, on the Colorado desert was found a specimen of Amaranthus ten feet high. A year later the same species in the same place matured in the drought at four inches. . . . Very fertile are the desert plants in expedients to prevent evaporation, turning their foliage edgewise toward the sun, growing silky hairs, exuding viscous gum."

Austin didn't join Coville in excluding Darwinian conflict from desert botany. She thought plants grew apart not just because of environmental restrictions but because they compete for water: "There is neither poverty of soil nor species to account for the sparseness of desert growth, but simply that each plant requires more room. So much earth must be preempted to extract so much moisture. The real struggle for existence, the real brain of the plant, is underground; above there is room for a rounded perfect growth." In the desert overall, Austin saw a hierarchy of competition: plants struggling with herbivores; herbivores with carnivores; scavengers poised to consume losers. It was just that aridity scattered the contenders and made the struggle inconspicuous: "So wide is the range of the scavengers that it is never safe to say, eyewitness to the contrary, that there are few or many in such a place."

Austin's affection for desert generated a large and patient curiosity about its inhabitants. She was an extraordinary observer of wild birds and mammals: "Watch a coyote come out of his lair and cast about in his mind where he will go for his daily killing. You cannot very well tell what decides him, but very easily what he has decided. He trots or breaks into short gallops, with very perceptible pauses to look up and about at landmarks, alters his tack a little, looking forward and back to steer his proper course."

Yet Austin's mysticism seemed to obscure her naturalist side when it came to origins and causes: "If one is inclined to wonder at first how so many dwellers came to be in the loneliest land that ever came out of God's hands, what they do there and why stay, one does not wonder so much after having lived there." She did not wonder in print about dried lakebeds and fossils. And much of the wildlife behavior she described so acutely was not especially desert adapted. She perceived some desert creatures' cleverness in finding water to drink but not others' independence from drinking. She found the latter idea suspect: "It is the opinion of many wise and busy people that the hill-folk pass the ten-month interval between the end and renewal of winter rains, with no drink; but your true idler, with days and nights to spend beside the water trails, will not subscribe to it."

Austin might not have believed that a lizard can live on creosote bush and wildflowers without drinking. Aside from noting that desert Indians all ate chuckwallas' "delicate white flesh," she covered lacertilians in a paragraph:

> There are myriads of lizards on the mesa, little gray darts, or larger salmon-sided ones that may be found swallowing their skins in the safety of a prickle-bush in early spring. Now and then a palm's breadth

of the trail gathers itself together and scurries off with a little rustle under the brush, to resolve itself into sand again. This is pure witch-craft. If you succeed in catching it in transit, it loses its power and becomes a flat, horned, toad like creature, horrid-looking and harm-less, of the color of the soil; and the curio-dealer will give you two bits for it, to stuff.

For Austin, the desert's magic lay mainly in the immediate human relationship to it. No enthusiast of Spencer's progress as it steamrolled the Owens Valley Paiutes she admired, she tacitly accepted its assump-tion that older forms of life must give way to newer ones, as with the "herb-eating, bony-cased old tortoise that pokes cheerfully about those borders some thousands of years beyond his proper epoch." Her curios-ity did not extend to learning just how different desert life might be from other kinds, or how the differences came about. Perhaps she feared such knowledge might obstruct her wishful self-identification with a tawny, voluptuous Sphinx.

Her main literary rival was more distanced from the desert but also more attuned to its past. John C. Van Dyke, a patrician professor of art history at Rutgers College in New Jersey, came to the desert for a lung disorder but stayed to admire its aesthetic qualities in a significant way. Van Dyke's firsthand knowledge of desert was shallower than Austin's. Some of his natural history was dubious, as was his claim to have explored California's Sonoran Desert from the coast to Mexico alone on horseback, living off the land. His writing is short on the logistics of such a feat. But however he managed it, Van Dyke saw a lot of desert and made provocative observations of it in his 1901 book, *The Desert,* which blandished readers with gorgeous atmospheric effects set against lurid desolation. It was a catchy formula at a time when Spencer's "survival of the fittest" jostled in salons with the "art for art's sake" of fin de siècle aesthetes, and since Van Dyke stood higher in the cultural mainstream than the semibohemian Austin, his work did more to shape contempo-rary attitudes:

The afternoon sun is driving its rays through the passes like the sharp-cut rays of search-lights, and the shadows of the mountains are lengthening in distorted silhouette upon the sands below. Yet still the San Bernardino Range, leading off southeast to the Colorado River, is glittering with sunlight at every peak. You are above it and can see over its crests in any direction. The vast sweep of the Mojave lies to the north; the Colorado with its old sea-bed lies to the south.

It is a gaunt land of splintered peaks, torn valleys, and hot skies. And at every step there is the suggestion of the fierce, the defiant, the defensive. Everything within its borders seems fighting to maintain itself against destroying forces.

Van Dyke contemplated the desert's past and the prevailing ignorance about it with similar elegance:

Everywhere you meet with the dry lake-bed—its flat surface devoid of life and often glimmering white with salt. These beds are no doubt of recent origin geologically, and were never more than catch-basins of surface water; but long before ever they were brought forth the whole area of the desert was under sea. Today one may find on the high table lands sea-shells in abundance. . . . The corals, barnacles, dried sponge forms, and cellular rocks do not differ from those in the Gulf of California. The change from sea to shore, and from shore to table land and mountain, no doubt took place very slowly. Just how many centuries ago who shall say? Geologists may guess and laymen may doubt, but the Keeper of the Seals says nothing.

Van Dyke took advantage of the desert's Sphinx-like silence to speculate engagingly, if confusedly, on its possible origins. Too sophisticated to ascribe them even perfunctorily to the "hands of God," he was too much the aesthete for stark materialism. So he tacitly invoked two origins.

One was creationist, although it involved a pantheist matriarch instead of a biblical patriarch: "Nature goes calmly on with her projects. She works not for man's enjoyment but for her own satisfaction and her own glory. She made the fat lands of the earth with all their fruits and flowers and foliage; and with no less care she made the desert with its sands and cacti. She intended that each should remain as she made it." Van Dyke clearly had read John Muir's rapt evocations of originative phenomena such as glacier-forming storms: "tender snow-flowers noiselessly falling through unnumbered centuries . . . messengers sent down to work in the mountain mines on errands of divine love." In his creationist mode, he sounded like a desert Muir: "I need not now argue beauty in the birds, the beetles, and the butterflies. . . . But why are not the coyote and the lizard beautiful too? . . . Nature's work is all of it good, all of it purposeful, all of it wonderful, all of it beautiful."

Pantheist creationism accounted for the serenity Van Dyke saw around his head. It did not account for the scramble for survival he saw around his feet: "And always here in the desert the question comes up: Why should Nature give these shrubs and plants such powers of endurance and resistance, and then surround them by heat, drouth, and the attacks of desert animals? It is existence for a day, but sooner or later the growth goes down and is beaten into dust." To explain this, Van Dyke invoked a Darwinian stepmother who had forced competition into a world created to be harmonious:

> The life of the desert lives only by virtue of adapting itself to the conditions of the desert. Nature does not bend the elements to favor the plants and animals; she makes the plants and the animals do the bending. . . . And so it happens that those things that can live in the desert become stamped over time with a peculiar desert character. The struggle seems to develop in them special characteristics and make them, not different from their kind, but more positive, more insistent. The yucca of the Mojave is the yucca of New Mexico and Old Mexico, but hardier. . . . Perhaps there never was a life so nurtured in violence, so tutored in attack and defense as this. The warfare is continuous from the birth to the death. Everything must fight, fly, feint or use poison, and every slayer eventually becomes a victim. What a murderous brood for Nature to bring forth!

Van Dyke's ambivalence inhibited him from Muir's firm advocacy of nature preservation, although *The Desert* does call for protection: "The deserts should never be reclaimed. They are the breathing-spaces of the west and should be preserved forever." But, with his lung problems, he may have been more concerned with preserving air quality than flora and fauna. Anyway, his book soon moves on to other matters, and he didn't lobby for wilderness as Muir did.

When the two writers met, they clashed. Muir was a friend of Van Dyke's brother, Theodore, who had a ranch in the Mojave. When his daughter had lung trouble, Muir sent her to stay there, and happened to visit her during one of brother John's visits. Theodore's son Dix recalled that they "wrangled incessantly about various highbrowed topics." The wrangling's cause is unclear, but since *The Desert* was published by then, Muir may have denounced its fashionable emphasis on the Darwinian "struggle for existence," a concept he disliked. Dix noted that, whereas Muir retreated "in great disgust" from a mass slaughter of jackrabbits attracted to the ranch's alfalfa fields, his uncle "derived some gratification from clubbing some."

Darwinist struggle evidently prevailed over pantheist mysticism in Van Dyke's everyday outlook. Yet, although more willing than Mary Austin to think about desert life's evolution, he seemed little clearer about it. He didn't, like her, deny extreme adaptations like the ability to live without drinking. "The desert animals seem to fit their environment a little snugger than either plant or human," he wrote. "For, strange as it may appear, many of them get no water at all." But he took non-drinking for granted as part of desert life's "endurance and resistance" without wondering how it came about, except that he doubted mammals and birds could get their water simply by eating dry desert plants. About plant-eating lizards, he didn't wonder at all, possibly unaware of them. Of reptiles in general he wrote, with characteristic hyperbole: "It would seem as though Nature had brought them into the desert only half made-up—a prey to every beast and bird. But no, they are given the most deadly weapon of defense of all—poison. Almost all the reptiles have poison about them in fang or sting."

Van Dyke's overall concept of adaptation was typically ambivalent, as with his notion that nature gives plants their "powers of endurance and resistance" and then "surround[s] them with heat, drouth, and the attacks of desert animals." In the usual evolutionary sequence, of course, the "powers of endurance" come *after* the "heat and drouth." Van Dyke followed the usual sequence elsewhere, as when he wrote that nature does not "bend the elements to favor the plants and animals" but makes the plants and animals "do the bending." The inconsistency evidently didn't bother him, as though he liked the intellectual exercise of wandering in circles.

Between them, Austin and Van Dyke left a heavy conceptual baggage for their successor, Joseph Chase. Her Sphinx metaphor, published a decade before his *California Desert Trails,* may have influenced Chase's. "It is as if you were bemused by the gaze of a sorceress," he wrote of the desert's allure, "or had listened over long to some witching, monotonous strain." But he didn't mention Austin, and he was ambivalent about sensuous mystiques, as though mindful that the Greek Sphinx was said to devour men in a sexual embrace. "Certainly it is not love, in any degree, that one feels for the desert," he continued, "nor could any other single term convey the sentiment."

Chase drew more from Van Dyke's mélange of aestheticism and Darwinism. After cataloguing the desert's ugly aspects in his preface, he added, "One feature of loveliness the desert has, however, that is essential: in one field of beauty it is supreme. That is the field of color." He based this on the authority of "Professor John C. Van Dyke, who has made that fine study of the desert which takes the rank of a classic." Chase's observations often sound like *The Desert:* "I do not see how Sahara, Gobi, or Arabia could improve on this for rigid nakedness and sterility. One here sees Mother Earth scalped, flayed, and stripped to the skeleton. Yet there is a strange beauty in it all. Perhaps the dormant savage in the breast, some strain of the Paleozoic, wakes up in the presence of these chaotic, barbaric shapes. I felt a sort of excitement, a half sense of recognition, as if something nudged and whispered—'Your primal home. Come back.'"

Like Van Dyke, Chase chose spiky plants such as cactus as "the special offspring of the desert," born from Darwinian competition: "With ingenious pains, Nature has wrought out this unique family, fitted to endure the very reverse of ordinary plant conditions. . . . Since leaves yield too much to evaporation, spines and thorns are adopted. Rainfall being a matter of doubt, the cactus models itself on the canteen, and fills up to the limit when it gets the chance. And since a canteen is a temptation to thirsty tramps such as jack-rabbit and coyote, the spines are hooked, barbed, clawed, and made as generally troublesome as possible."

A former church social worker, Chase was even more ambivalent about nature than Van Dyke. Writing for tourists, he sweetened his Darwinism with muscular Christianity and rationalized the desert's harshness as a manifestation of divine grandeur: "With all my weariness, I do not think I have ever been so charmed as that evening by the sunset coloring. . . . It was more than Nature, infinitely more than Aesthetics. Some words of the Psalter came to my mind—'Who deckest Thyself with light as it were with a garment.' Yes, only that expressed it: it was the Vesture of God." But having actually struggled through it on horseback as his narrative shows, Chase also saw the desert less piously as a treacherous foe that had threatened him in painfully degrading ways. This colored his attitude to its organisms. He killed spiders, scorpions, and snakes whether they endangered him or not, and he "experimentally" tortured a sidewinder to death by confining it in the midday sun. Even harmless lizards were "bony little goblins with sharp tails and a leer in the eye that comes near being devilish."

For all his animosity to desert, Chase came closer to protecting some than Austin or Van Dyke. He advocated a national park in the canyons around Palm Springs, and the idea caught on. In 1922, Congress approved a 1,600-acre national monument there. But the Cahuillas, who figure in Chase's books as the Paiutes do in Austin's, had veto power because it was on their reservation: "To the astonishment of everybody, the Indians vetoed the bill. Moreover, to make certain that tourists understood whose land it was, they erected a tollgate, built a road, and started charging admission."

Anyway, Chase's park was for palms and springs, not sidewinders and cactuses: "There are, it is true, about the fringes of the desert, spots of sylvan beauty. . . . But these are only local conditions, quite the reverse of typical . . . the desert yields no point of sympathy, and meets every need of man with a cold, repelling No." The idea of protecting broad swathes of the blazing scrub that had tried to kill him probably would have seemed bizarre.

Yet if the repellent, hypercompetitive desert that Van Dyke and Chase perceived beneath its colorful surfaces made for tepid conservationism, it also threw a cold light on popular notions of progress. Van Dyke had a bleak sense of the desert's implications for those:

It has been swept by seas, shattered by earthquakes and volcanoes, beaten by winds and sands, and scorched by suns. Yet in spite of all it has endured. . . . And yet in the fullness of time Nature designs that this waste and all of the earth with it shall perish. Individual, type, and species all shall pass away; and the globe itself become as desert sand blown hither and yon through space. . . . And how came it to die? What was the element that failed—fire, water, or atmosphere? Perhaps it was water. Perhaps it died through thousands of years with the slow evaporation of moisture, and the slow growth of the—desert.

Van Dyke implied that Darwinian competition, carried to logical extremes in places like the desert, eventually might produce not the progress that Herbert Spencer anticipated but a planet increasingly hostile to life. Not merely "reclaimable" wastelands, such places might be evolutionary lesions spreading relentlessly through the biosphere. Darwin had similar thoughts in his darker moods. "What a book a Devil's chaplain might write," he exclaimed in a letter to his botanist friend, Joseph Hooker, "on the clumsy, wasteful, blundering low & horribly cruel works of nature!"

Apocalyptic Darwinism, however, still skimped on the details of exactly how the desert produced its brood. It would take more than ideology, mysticism, aestheticism, or even close observation to uncover some of those. And the details that were uncovered would show a surprising propensity to shift around according to the various angles from which they were seen.

NINE

Hopeful Monsters

Early twentieth-century scientists forgot Edward Cope's neo-Lamarckian ideas, but they still faulted Darwin for not explaining "the origin of the fittest," the source of the variation that he said natural selection needs to operate. Without such a source, evolutionary change of any kind would not occur, since organisms would simply remain the same from generation to generation. Many evolutionists became so obsessed with variation's origin that they almost forgot about Darwin. It seemed possible that changes in the "germ plasm," genetic mutation, might be the main force in evolution, with natural selection a secondary factor. Mutations might allow organisms to give birth to new forms without undergoing a lengthy, random process of "descent with modification," as Darwin originally called evolution. They might produce "hopeful monsters," offspring that differed radically from their parents in ways that luckily made them much more fit to survive and reproduce in their turn.

An early attempt to explain California desert evolution reflected this. In 1902, the Carnegie Institution of Washington, D.C., established a desert research laboratory in Tucson. Its first director was Daniel Trembly MacDougal, originally of the New York Botanical Garden. Desert biology might seem a stretch from that genteel setting, but MacDougal was a sportsman and socialite as well as a botanist. He divided his time between the Tucson lab, a Carmel, California, home (where he became an intimate friend of Mary Austin in her postdesert years), New York and Washington clubs, and field expeditions that included research in the Colorado Delta region. In the process, he acquired a wide if not necessarily deep acquaintance with biological issues concerning the desert's past: "Desert conditions seem to have prevailed in . . . southwestern America since Cretaceous times. In the . . . Salton basin of California . . . the characteristic vegetation is composed largely of spinose and switch-like forms in which the chief development has been toward restriction and induration [hardening] of surfaces; a result attributable to the degree of aridity, the seasonal distribution of the rainfall, and also to the intervention of great climatic oscillations."

When MacDougal wrote this, the Cretaceous period, the late dinosaur age, was still thought to have ended a few million years ago in accordance with Lord Kelvin's hundred-million-year-old planet. If desert plants had evolved in a few million years, the gradual Darwinian process seemed inadequate to explain them:

> The view that such forms are of recent origin, that is since the Cretaceous Period within the present period of advancing desiccation, would predicate a very great phylogenetic activity.

> Adaptation, therefore, furnishes but an insecure basis upon which to found a theory of the origin and development of any flora, inclusive of those inhabiting arid regions. The influence of external conditions upon the germ plasm, however, has been seen to produce irreversible changes in a hereditary line by which new combinations of qualities and new characters were called out, which were fully transmissible.

MacDougal inferred that desert plants were the product of environmentally induced genetic mutations: "Gradual modifications by which a long series of forms, each slightly different than its immediate progenitors, appear to have been found among animals, but with plants no such series has been brought to light. These organisms, on the contrary, exhibit sports or saltatory [sudden] derivatives which now have been seen and recognized in a number of species. Such mutants are now occurring, and we may predicate with certainty that they have occurred with normal frequency during the formation of the deserts of southwest America."

So, instead of invading from older deserts to the south as drying climate drove Asa Gray's lush Arcto-Tertiary flora north, the ancestors of California's cactuses and other desert plants might have already lived in California, where they could have mutated rapidly to adapt to the new conditions. A better manifestation of the "hopeful monster" idea than cactuses would be hard to imagine, from a cactus's viewpoint, at least.

The problem with this scenario was that fossils of the presumed ancestral plants were unknown. MacDougal was aware that most fossils in California desert are of organisms that lived in ancient seas or lakes rather than dry land. Still, he didn't think that precluded desert origins within the region. Desert plant fossils could be expected to turn up eventually: "It is true, of course, that desert conditions are not favorable for fossilization, yet many opportunities for such action undoubtedly occur in the carrying and burying action of the torrential floods of desert streamways while wind blown deposits might preserve the more indurated forms. Many of these and the skeletons of the Cactaceae would seem well adapted to preservation in this manner."

Yet when plentiful land fossils did turn up in the Mojave two years after MacDougal proposed his mutant desert genesis, they weren't what he'd anticipated. In 1911, geology students who had been studying strata around Barstow and Red Rock Canyon began bringing ancient bones to John Merriam, the first professor of paleontology at UC Berkeley. Not to be confused with the Death Valley Expedition's C. Hart Merriam, he also explored California's remotest places, but for extinct instead of living fauna. He'd already found some of the oldest known fossil reptiles in the Klamath Mountains. In 1915, after excavating more of the California desert bones, Professor Merriam wrote: "With the exception of the John Day region of eastern Oregon, the largest part of our knowledge of mammalian life west of the Wasatch [the Rockies] is obtained in the heretofore unexplored deposits of the Mohave Desert. At the present time, there are available from the Mohave at least three extinct mammal faunas previously unknown, or only imperfectly known, in the Great Basin."

The faunas dated from the Miocene epoch, halfway between the dinosaur age and the present, and from the subsequent Pliocene epoch just before the ice age, so they might have been expected to throw at least some light on desert origins. They included some animals that inhabit California desert now—rabbits, pronghorns, foxes, tortoises—and a few that live in deserts elsewhere—camels and peccaries. But they also included creatures that would be bizarre in today's Mojave—mastodons, at least four horse species, saber-toothed cats, large dogs, and extinct ungulates called oreodonts. It was confusing: "In a few strata abundant remains of fresh-water mollusks indicate deposition of these beds in fresh-water ponds or lakes. At other levels the skeletons of large desert tortoises and numerous remains of land mammals now characteristic of flat open country suggest accumulation upon dry land."

Some of Merriam's colleagues thought the fossil faunas could have inhabited a desert. But he doubted it, bearing in mind the prevailing idea that the West Coast's mountains began to rise and form a "rain shadow" by cutting off oceanic precipitation only in the late Pliocene epoch. This implied that southeast California had been more low lying and level before, more like today's Great Plains or other grasslands. In any case, Merriam found no fossils of cactuses or other desert plants buried with the mastodons and horses. "As nearly as the writer can judge," he wrote, "the climate conditions in the Mohave area through the period in which the mammal beds were being laid down, were those of a semi-arid region somewhat more humid than the Mohave today, and the climate corresponded approximately to that now obtaining in the southern end of the Great Valley of California."

It seemed that William Manly and John Fremont had discovered the

Mojave too late. According to Merriam, it once had been just such a pastoral paradise as the pioneers craved. True, his reports on the Miocene and Pliocene faunas included no mention of grass fossils—or any plant fossils to speak of. Merriam complained that they had found no fossil sites where ancient streams had carried material down from adjacent hills. Since vegetation is often richer on highlands than plains, that might have provided more information on the flora. But the Sphinx, having released the tantalizing bones, resumed a retentive silence.

TEN An Old Earth-Feature

An anti-Darwinian bias toward desert evolution lingered well into the twentieth century. Forrest Shreve, a botanist who succeeded Daniel MacDougal as director of the Carnegie desert lab, published a book about cactus in 1931. It reiterated Reverend George Henslow's theories of the 1890s:

> One of the most commonly held ideas about the cacti is that their spines have come about 'for the purpose of protecting them' from their enemies in the animal kingdom. True enough, there can be little doubt that the spines do the plants good service in many cases in protecting them from the ravages of rodents and larger animals. That spines became characteristic of the whole cactus family because of the operation of natural selection is very doubtful indeed.
>
> It would be difficult to find a competent biologist who would main-

tain that the less spiny plants were eaten and the more spiny ones protected and preserved, and that for these reasons the members of the cactus family gradually procured their characteristic spininess. It is now held to be far more probable that the development of spines is a direct physiological effect of the dry atmosphere and scanty water supply.

F. B. Sumner, a zoologist at the Scripps Institution in La Jolla, thought the pallor of desert animals, which Darwin had seen as protective coloration, might be another example of direct environmental influence. "There are strong reasons for doubting whether this prevalent pallor of desert animals has been developed owing to its effectiveness for concealment, and there are some reasons for attributing it to the direct or indirect effect of climatic conditions upon pigment formation." Sumner poked mild fun at popular Darwinism and its murderous brood:

> Great stress has been laid by some writers upon the supposed intensity of the struggle for existence in the desert. This conception has found brilliant literary expression through the pen of J. C. Van Dyke ('02) [sic] whose charming little volume, *The Desert*, contains so much of truth, as well as beauty, that it seems heartless to point out its scientific inaccuracies. . . . Is it really true that the beasts and birds are fiercer or more cunning in desert than elsewhere? Is it true that we meet with a larger proportion of venomous snakes or spiders or insects? These things are possible, but we cannot accept them merely because they appeal to our sense of the picturesque.

Dismissing Darwin was getting harder, however. Sumner noted that experiments had shown that desert animals' pallor was inherited, which worked against the environmental influence theory. And he had doubts about environmentally induced cactus spines: "Despite arguments to the contrary, there are reasons to believe that the spines of some desert plants serve as defensive armament, and that this need for defense has been responsible in part for their development." Natural selection was coming into vogue again.

The discovery in 1906 by another aristocratic English physicist, Lord Rayleigh, that radioactivity continually renews the earth's internal heat had challenged Lord Kelvin's hundred-million-year-old planet. Estimates of the time elapsed since the dinosaur age gradually lengthened toward the presently accepted sixty-five million years. This allowed more leeway for natural selection to work, and the fossil record kept revealing more of the slow changes that selection might produce. At the same time, genetic discoveries made it seem less likely that environmentally induced characters could be inheritable or that mutation could spontaneously produce a highly modified organism like a cactus. Laboratory studies of fruit flies showed that most mutations are neutral or harmful, suggesting that more than genetic change is needed for "the origin of the fittest."

C. Hart Merriam, who had described lizard speed and greed in the Death Valley Expedition report, came down hard on hopeful monsters: "The question before us is, Do species among the higher vertebrates originate by a sudden acquirement of new characters? In seeking the answer, I have passed in review more than a thousand species and subspecies of North American mammals and birds without finding a single one that seems to have originated in this way." Measuring kangaroo rat legs and tails, Merriam found a small, constant rate of variation. "What does this mean?" he asked. "It means . . . that all the species of the genus *Dipodomys* have come to a halt along a common line, like soldiers in a well-drilled regiment indicating that in the course of their evolution from a generalized to a specialized type, they have already reached . . . a state of equilibrium and equipoise, from which any marked departure is injurious if not fatal. . . . These animals have numerous enemies and a multitude of competitors, which means that the struggle for existence is always severe."

In 1936, the doyen of American plant ecologists, Frederic E. Clements, published an article on desert origins typical of the mainstream's drift back toward Darwinism. Clements had pioneered the concept of plant succession while watching abandoned Nebraska farm fields revert from crops to weeds and finally back to what he called a "climax vegetation" of native prairie grasses and forbs. He likened climax vegetations (in effect, ecosystems) to organisms, speculating that they also "evolve" through natural selection of favorable traits and competition with other climax vegetations. He interpreted creosote bush, burroweed, and their scrubby associates as a "desert climax," originating in Mexico, that had competitively replaced earlier plant climaxes after rising coastal mountains blocked precipitation in the late Miocene and Pliocene. The earlier vegetation had first been Arcto-Tertiary forest, as Asa Gray had maintained, then savanna and grassland as John Merriam's horses and camels seemed to show.

Even Forrest Shreve, despite his anti-Darwinian attitude about cactus thorns, was inclined to accept natural selection and competition in desert evolution overall. "There is evidence that the area concerned has been desert at least since the Miocene," he wrote, "and that it has been the arena of prolonged combat in which a considerable number of plant families have contributed members able to persist in favorable spots or at suitable seasons."

Still, as a field botanist who ran the Carnegie lab until 1945, Shreve doubted that fossils could reveal the Sphinx's deepest secrets. "There appears to be little hope that the paleontological record will give much aid in reading the history of the origin of the present distinctly desert plants of this region." He placed his hopes on more field botany: "It is believed that a study of the floristic features, interpreted in terms of the ecological behavior of the plants involved, will do as much as can be done toward unraveling the biological history of the area and its relations to the development of similar plant populations in northern central Mexico and in South America."

A younger botanist was happily pursuing such a study while Shreve endured the dust, heat, and administrative burdens of the Tucson lab. Born in Los Angeles in 1898, Ivan M. Johnston had earned his doctorate at UC Berkeley and had collected plants in Baja, before moving east. He'd worked with Clements at the Carnegie Institution and then had ascended to Harvard as an assistant at the Gray Herbarium and a research associate at the Arnold Arboretum. He'd continued to specialize in desert plants, traveling to Mexico and the Southwest frequently, and in the mid-1920s he made a solo expedition to South America that became legendary to fellow botanists like Richard A. Howard, the Arnold Arboretum's director: "His field work in Chile was both energetic and successful. Physically active and personable, he made many friends in the boat trip south and received invitations to remote areas for his collecting. His interest in . . . entering into new areas, led him from the driest of desert areas to Andean locations at 17,000 feet. Many of his collecting locations have not been revisited as they represent remote locations difficult to reach."

In 1940, Johnston synthesized his work in a paper that became a touchstone of the field botany version of desert origins. Comparing the plants of North and South American deserts, now isolated from each other, Johnston found numerous similarities that suggested ancient relationships. Creosote bush, although so common in North America, was

> clearly a South American type, for it has several congeners in the Argentine deserts and its family, the Zygophyllaceae, a world-wide group of chiefly desert shrubs, has one of its present centers there. . . .
>
> Two other shrubs occur in indistinguishable forms in the deserts of both North and South America. *Atamisquea emarginata* [a leafy member of the caper family, unattractively called vomitbush] is a companion of *Larrea* in the dry Monte of western Argentina and with *Larrea* again it is present in a much more limited area about the Gulf of California in northwestern Mexico. And again there is *Koeberlinia spinosa* [a nearly leafless caper family member known as crown of thorns] widely distributed, though not particularly common, in the deserts of northern Mexico and adjacent United States, and also present in a limited area of the dry Chaco of Bolivia.

Johnston thought that desert adaptations such as leaf reduction, spines, and resins were more prevalent in South American plants:

> When it is realized that most of the North American species showing such modifications, as the glutinous varnish and the leafless habit, are members of genera also represented in South America, where we know that these types of modifications are frequent and characteristic, we must be prepared to admit that, at least in many instances, these habits must have spread to North America. These are old habits of a flora now characteristic of South America. . . . That these modifications are present in various northern members of the genera, whose species no longer remain identical or very similar on the two continents, is simply evidence of a long period of diversification following the time when American desert shrubs were exchanged between the continents.

Johnston's observations supported Darwinist gradualism in suggesting that desert organisms had evolved over a long period:

> Since most biologists appear to think of geological climates only in terms of ice-ages and wet, usually tropical conditions, perhaps I should emphasize the fact that deserts are an old earth-feature. The world must have always had its deserts, at least, those just outside the tropics. There has always been moist ascending air, and rain, near the equator, and descending dry air, and aridity, at about latitude thirty. Desert floras may well have an age and continuity comparable to the floras of the wet tropics. Many groups of plants such as the Zygophyllaceae and Chenopodiaceae have probably been evolving in deserts, at least since Mesozoic time. And these may be relatively recent xerophytes as compared to *Ephedra* and *Welwitschia*.

The fact that the desert gymnosperm, *Ephedra*, grows in America and another desert gymnosperm, *Welwitschia*, grows in Africa was yet another suggestion that deserts might be very ancient. Johnston cited examples of other closely related desert plants that live in both Africa and the Americas. These include *Thamnosma*, turpentine broom, the purple-flowered angiosperm shrub that I mistook for *Ephedra* on my first trip to the Providence Mountains. It has South African relatives. Compared to redwoods, of course, these "switch-like" shrubs aren't very impressive "old earth-features." But they are here, in vast numbers.

Johnston's study didn't show exactly when or how desert had come to occupy California, but it did give the inquiry a certain sense of direction. It must have seemed to field botanists like Shreve that desert biology would finally stop going in circles. But paleobotany was about to throw them a curve.

ELEVEN

A Climatic Accident

Ivan Johnston presented his desert shrub paper at the Eighth American Scientific Congress in Washington, D.C., in May 1940, and it was influential. A 1944 textbook resoundingly titled *Foundations of Plant Geography* echoed its idea that very ancient South American deserts had been the source of ancient North American ones. But at least one member of the audience may have questioned Johnston's version of the desert past.

Daniel Axelrod was working as a postdoctoral fellow in paleobotany at the National Museum and the Carnegie Institution when Johnston presented his paper. The subject interested him, albeit from a different angle. When Axelrod was getting a PhD in paleobotany at UC Berkeley during the Depression, most junior scientists couldn't afford to do much larking off on fieldwork such as Johnston had enjoyed in the 1920s. But

Axelrod was an unusually energetic paleobotanist—probably one of the most energetic ever—and in 1940 he'd already been investigating the California desert's past for six years.

Energy ran in the family. Having emigrated from Russia to Brooklyn, where Daniel was born in 1910, the Axelrods hurried on to Guam, of all places. Then they moved to Hawaii, where young Daniel became a surfer, then to the San Francisco Bay Area, where he joined the Boy Scouts and took up hiking. This led to an interest in forestry, and Axelrod put himself through the UC Berkeley Botany Department by working on Forest Service plant surveys in the summer. But forestry and field botany didn't offer enough intellectual stimulation for a student who was avidly absorbing geology, climatology, and evolutionary biology as well.

Displaying characteristic decisiveness, Axelrod recalled an encounter that did offer enough stimulation: "In my junior year, Harry D. MacGinitie, then an 'old man' of about 35–40 years, came to the class in Forest Botany carrying a tray with well-preserved fossil plants—the Trout Creek flora. I asked what he was doing with them and learned that he was going to the herbarium to identify the species. I told him I knew what several of them were, so we went to the herbarium and looked them up—I was correct. I then realized that this was the sort of thing I would like to do." Axelrod didn't specify whether the sort of thing he would like to do was identify fossils or be proved correct about them. Doubtless it was a bit of both. He did develop very strong opinions on fossils and evolution during a career that would last the rest of the century.

California desert fossils puzzled him for several years, however. His early work tended to support the field botany version developed by Frederic Clements, Forrest Shreve, and Ivan Johnston. *Foundations of Plant Geography* noted this: "Fossil evidence bearing directly on the problem of the origin of the desert climax is not very abundant, although recent Pliocene and Pleistocene discoveries by Axelrod indicate the Mexican origin of the vegetation and largely substantiate Clements's account." Axelrod's first publication on a California fossil flora, from the several-million-year-old Pliocene Eden Beds in Riverside County, implied a respectable antiquity for at least some local desert vegetation. Although the fossils mainly consisted of plants typical of nondesert southern California today—manzanita, ceanothus, live oak, Coulter pine, and ghost pine—they apparently included two taxa that now live in Mexican desert:

> The occurrence of a desert element comprising *Platanus* (cf. *P. wrightii*) and *Sapindus* is consistent with the southern occurrence of this fossil flora, and is indicative of more arid conditions during the Pliocene than exist at the fossil locality today. *Sapindus* and *Platanus* reach their best development in the desert under a rainfall of 5 to 10 inches annually, while the chaparral now growing at Beaumont exists under a rainfall of approximately 20 inches. . . . With increased rainfall in the late Pliocene and Pleistocene, the desert elements probably retreated southward along available lines of migration, since these species are no longer found in California.

But *Platanus* and *Sapindus*—sycamore and a small tree called soap-berry—are riparian plants in desert: most sycamore and soapberry species occur in woodland. When Axelrod published a longer piece on the Eden Beds a few years later, the picture had become more complex. The sycamore in the Eden Beds had begun to seem more like the species in nondesert coastal California today, *Platanus racemosa*. And although fossils of modern desert plants like mesquite, ephedra, and a sunflower family shrub called scale broom (*Lepidospartum*) occurred in the Eden Beds, Axelrod didn't find enough of them to confirm his previous impression: "True desert conditions such as are found in the present adjacent Colorado Desert may not have been existent in the basin during Mount Eden time."

Axelrod wrote his doctoral thesis on another fossil flora located just west of the present Mojave, in the mountains above Tehachapi Pass. The Tehachapi flora of fossil leaves, fruit, and wood that he dug from white ash deposits in volcanic beds was of Miocene age, millions of years older than the Eden flora. According to Johnston's old desert paradigm, it might have been expected to contain many desert elements. It did have what Axelrod called "a typical desert scrub association" of desert peach *(Prunus)*, mesquite *(Prosopis)*, and a spiny buckthorn relative called abrojo *(Condalia)*. But he thought the scrub had grown only on dry lower slopes. The overall vegetation was lusher, with subtropical trees like bursera and euphorbia bordering the scrub; ash, walnut, sabal palm, and cottonwood sharing riparian areas with sycamores; an oak savanna with ceanothus, buckthorn, and locust on uplands; and a woodland of pine, cypress, madrone, wax myrtle, and bay like that of today's coastal California on hills.

Axelrod found no fossil evidence in the Tehachapi site that creosote bush, ocotillo, cactus, or turpentine broom had grown there. Rather, the site suggested to him a much less mountainous landscape than today's, where twelve to twenty-five annual inches of rain watered woodland that had a dry season and semiarid patches but was otherwise unlike desert. He perceived similar conditions in other southern California fossil floras, including even older ones. A site from the Eocene epoch, twenty million years earlier than the Tehachapi flora, contained ancestors of walnut, wax myrtle, and soapberry.

Military service in the Pacific analyzing the aerial photos used to prepare for island invasions interrupted Axelrod's career from 1942 to 1946, although he kept publishing on fossil floras. Then he taught in the geology department at UCLA, which probably frustrated him by limiting his research time. Although a good lecturer, he was impatient with academic chores like acting as a thesis advisor. He said that few students knew enough to make it worthwhile. Within a few years, nevertheless, he managed to publish a ringing challenge to the established desert origin paradigm:

> A wide diversity of opinion exists with respect to the age and derivation of modern desert environments. According to one view, desert vegetation is an "earth-old feature." A second theory, corollary to this, is that desert floras of essentially modern character have been in existence in their present positions since angiosperms first assumed dominance during the Cretaceous Period. Although both opinions have been expressed on several occasions during the past decade, no evidence supporting them has been presented. . . . Analysis of the available data shows clearly that there were no desert environments of wide extent during those times. Thus a third opinion, and the one to be elaborated here, is that the desert environments now characterizing wide subcontinental regions are a phenomenon of only the latest part of geological time.

Axelrod claimed Daniel MacDougal as an intellectual forebear, citing his declaration that "xerophilous types of vegetation are of comparatively recent origin" and his belief that California desert plants had evolved here. He ignored MacDougal's ideas of sudden genetic mutation, however, and otherwise distanced himself from the previous scientific generation. Unconcerned with ideas of plant succession and ecosystem "evolution," he dismissed Frederic Clements's notion that a creosote bush and burroweed "desert climax" had invaded from Mexico recently, competitively replacing other climax vegetations. "Clements's postulate of a dominant hardwood-deciduous forest in the region now desert during Eocene and early Oligocene" was "untenable" and his "belief" that grassland dominated the region in the mid-Tertiary was "also in error."

Axelrod took longer to dispute Ivan Johnston's idea of an ancient South American desert spreading north in the dinosaur age, but he did a more thorough job:

> There appears to be no support from the geological record for the recently stated belief that "the desert floras may well have an age and continuity comparable with the floras of the wet tropics (Johnston, 1940)." In all cases known, their present areas were occupied by more mesic vegetation, not only in Tertiary time, but in Cretaceous as well . . . the present regional desert climates and flora seem to be no older than late Cenozoic anywhere in the world. Such environments are not "earthold features." They may more appropriately be called "climatic accidents" for they have been rare in earth history.

Axelrod's recent desert had one particular advantage. It was more compatible than an "earth-old" one with the long-prevailing idea that West Coast mountains like the Sierra Nevada rose within the past few million years, causing a desert-forming rain shadow. Newer evidence seemed to support this. Francois Matthes, a geologist who studied the Sierra in the 1920s, decided that high plateaus there are ancient relics of low-lying, level landscapes that have been elevated along with the peaks since Pliocene times.

To be sure, the more geologists studied the landscape, the more complicated it seemed. By the 1950s, they knew that high mountains had existed in the Sierra Nevada's present location during the mid to late dinosaur age. The granitic rocks of today's Sierra originated as molten magma when those mountains arose over two hundred million years ago. Most thought that the ancient mountains, called the Nevadan orogeny, had eroded away into plains and low hills by the beginning of the Age of Mammals, although there wasn't conclusive geological proof of this. Indeed, Axelrod regarded his nondesert plant fossils as some of the best available evidence that southeast California was more low lying, level, and moist than it is today for most of the past sixty-five million years.

Axelrod had to acknowledge Johnston's point that some long-distance migrations of species like creosote bush must have taken place, but he didn't think there had been enough of them to prove that the desert is old: "Such migrations account for only a small fraction of these floras, which are otherwise largely distinct." He maintained that most California desert species evolved in the past three to five million years from plants that lived in the preceding woodlands and savannas but were able to survive and adapt in the climatic accident of desert.

Axelrod cited the desert's most attractive feature, its spring wildflower displays, as one living proof of its youth:

> Increasing topographic and climatic diversity of these areas during Cenozoic time has resulted in the differentiation of numerous species (and varieties) well adapted to these more narrowly defined environments of the desert and adjacent regions. Thus it is not surprising to find that many desert herbs are polyploids [a mutation involving multiples of the normal diploid chromosome number] whose nearest ancestors have equivalents in the forest, woodland, scrub, and grassland vegetation now marginal to the desert region. The desert floras must therefore have evolved since their more mesic ancestral species inhabited the lowlands of the area now desert.

It is striking how many desert spring wildflowers are related closely to ones in the much lusher landscape of the coast. Of the species I've mentioned in places like Red Rock Canyon and the Providence Mountains—fiddlenecks, poppies, larkspurs, lupines, phacelias, paintbrushes—almost all have congeners common in coastal grassland and woodland. Most of the desert's wildflower species are endemic to it, but plants so much like coastal relatives might indeed have diverged from them recently.

Axelrod also cited desert shrub genera like *Baccharis, Rhus,* and *Prunus* that have close relatives (coyote bush, poison oak, and wild cherry) in coastal habitats. It was harder, however, for him to explain the majority of desert shrubs that have no close relations in coastal grasslands or woodlands. Many of those come from groups centered in the tropics. But he pointed out that many have more moisture-loving close relatives to the south in Mexico and Central America, and he maintained that they could have evolved their desert forms recently, like the wildflowers. His theory wasn't perfect, Axelrod seemed to say, it was just better than the others:

> The task of determining the origins of modern desert vegetation is by no means an easy one, as is amply evident from the preceding discussion. The problem is one whose solution rests largely with paleobotany, though it is clear that data from modern plant distribution can give important clues and that cytogenetic studies can aid greatly in determining centers of differentiation. The procedure followed above for interpreting the history of the Great Basin, Mohave, and Sonoran deserts appears sound. Furthermore, the general pattern of desert development elsewhere in the world seems to parallel closely that of western North America, with respect thus to age and origin.

Little published evidence exists of the botanical establishment's reaction to Axelrod's brusque dismissal of its ancient desert paradigm. That there was no reaction seems unlikely, especially given what he wrote about his opponents viewing desert as an "earth-old feature." This purple epithet is too much like Ivan Johnston's description of desert as an "old earth-feature" in his 1940 paper to be accidental. Axelrod may not have been mocking Johnston by reversing "old" and "earth." The 1944 textbook, *Foundations of Plant Geography,* reversed them solemnly: "The *Larrea-Fransera* desert developed through . . . immigration of Mexican Sonoran species from the south where the desert climate and climax have long existed because they are an earth-old feature." Still, such little things can prickle, like a cholla spine on a pants leg.

Johnston himself was increasingly hors de combat by the 1950s. Harvard's administrative politics reduced him to a distracted outsider, and he published infrequently before his death of a heart attack in 1962. But Axelrod definitely nettled others, setting a lifelong tone. He particularly nettled geologists, some of whom looked askance at his use of plant fossils to dictate the landscape's evolution over the past sixty-five million years. As a California colleague said four decades after Axelrod first challenged "earth-old" desert: "I thought by all means he should have been elected to the National Academy of Sciences, but he's a very prickly person, and he raises a lot of hackles. There were two or three members of the geology section of the National Academy with whom he had particular quarrels, and it became clear that no matter how much we pushed from our side that there was opposition so he would never get in."

TWELVE An Evolutionary Frontier

The colleague who recalled Daniel Axelrod's prickliness underwent some controversies of his own. He was G. Ledyard Stebbins, who was revolutionizing plant evolution in general while Axelrod revolutionized desert plant evolution in particular. As one of the chief proponents of the "new evolutionary synthesis," also known as neo-Darwinism, Stebbins helped to shoulder the contentious task of resurrecting natural selection from the limbo into which opposing ideas and its own original short-comings had thrust it.

By the late 1930s, discovery of basic genetic mechanisms had con-founded non-Darwinian notions about "the origin of the fittest." Whether or not heat and dryness could make a plant's leaves more spiny or leath-ery, the microscope showed that such characteristics were passed to offspring via dividing chromosomes. It was the genetic material, later

identified as DNA and RNA molecules, that produced inheritable variation, not the environment. On the other hand, genetic variation could
not escape the environment's influence. Spines or leathery leaves were
more likely to be inherited over time if they contributed to reproductive
fitness. Neo-Darwinism was a "new evolutionary synthesis" because,
through a number of scientific disciplines, it showed that genetic mutation is the source of the variation that natural selection requires instead of
the alternative to natural selection that early geneticists had envisioned.

By identifying the source of variation, neo-Darwinism changed the
implications of natural selection. Extending that concept to biological
categories above the individual organism, as with Frederic Clements's
"evolving" ecosystems, became harder. The bushes and lizards in an
ecosystem like desert evolve by natural selection because they have
genes: the ecosystem doesn't. An older biological term, "development,"
better describes how things like ecosystems change, although tractmongers unfortunately have co-opted the word. I sometimes refer to "desert
evolution" in describing ecosystem-level change because if I referred to
"desert development" readers would think of shopping malls instead of
bushes and lizards.

Neo-Darwinism's chief proponents were geneticists with an interest
in field biology or vice versa. A Russian émigré, Theodosius Dobzhansky,
was America's first effective proponent because he came from a tradition in which field and laboratory biology were not as alienated as
in the English-speaking world. His 1937 book, *Genetics and the Origin
of Species*, encouraged several young scientists to rethink Darwinism:
zoologist Ernst Mayr, paleontologist George Gaylord Simpson, botanist
G. Ledyard Stebbins.

In contrast to jaunty Axelrod, Stebbins was an unlikely revolutionist.
He came from an old New York family that was, as he remarked, "not
upper crust, but . . . getting near there," and that expected him to enter
law or finance. He duly attended prep school, but books about scientific
exploration and summers dabbling in natural history on Mount Desert
Island in Maine made him dream of a biology career. Eventually, he
convinced his surprised father that the subject could be respectable,
and he studied botany at Harvard. But Stebbins felt the same dearth of

stimulation in academic botany as Axelrod: he also moved on to chancier territory, not the physical rigors of fossil hunting but the cerebral ones of genetics, still avant-garde in the late 1920s. His Harvard PhD thesis, "The Cytology of *Antennaria* [pearly everlasting]," was considered so advanced that it had trouble getting past the graduate committee.

Stebbins taught at Colgate until 1935 and then moved to UC Berkeley to work with E. B. Babcock, a leading plant geneticist who was studying *Crepis,* a sunflower family wildflower, in a program that has been called the botanical counterpart of fruit fly genetics. Quick to breed and easy to keep in the laboratory, fruit flies had demonstrated the potentials and limitations of genetic mutation in animals. *Crepis* did much the same for plant mutation.

Stebbins's intellectual restlessness was better suited to California's fluid society than to the East Coast's more formal one. But he was still an odd man out even at Berkeley. The Botany Department was not interested in his preference for counting chromosomes over describing and naming species, while the lab specialists who ruled genetics didn't share his fascination with the living world's evolutionary problems.

Stebbins's eccentricities generated new thinking on desert origins, however. Whereas Darwin had seen desert as an evolutionary backwater where aridity minimizes natural selection, Stebbins began to see it as an evolutionary frontier where aridity encourages selection. Darwin thought that desert mutes change by discouraging competition among organisms in favor of static adaptation to environment. Stebbins thought it could amplify change by providing unusual opportunities for organisms to express genetic variability. California was a good place for testing such ideas, although Stebbins first concentrated on the coast's semiarid Mediterranean flora.

"Rapid evolution requires the presence of rapidly changing secular [breeding] environment," he wrote in a 1947 paper. "But not all plant groups exposed to such changes can respond to them by evolving. There must be present in the group not only genetic variability, but in addition gene combinations which are preadapted to the direction of the secular change. Thus in coastal California in recent geological time, rapid evolution has been chiefly in annuals, bulbous perennials, sclerophyllous shrubs, and other plants adapted to drought and highly seasonal precipitation." The fact that many such California plants have evolved dozens of endemic species—annuals like phacelia and meadow-foam, bulbous perennials like mariposa tulip and brodiaea, arid-adapted shrubs like manzanita and ceanothus—seemed evidence of rapid evolution.

Stebbins incorporated such ideas into his book, *Variation and Evolution in Plants,* which made him one of neo-Darwinism's leaders on its publication in 1951. The editors of his collected papers affirmed its continued importance in 2004: "The book, which synthesized perspectives on plant evolution from a range of disciplines including plant genetics, systematics, population biology, paleobotany, and plant geography, is generally regarded as the synthetic work which brought varied botanical fields into the intellectual event called the evolutionary synthesis. . . . He stressed the centrality of natural selection as the dominant mechanism during evolutionary change, but also left plenty of room for random genetic drift and non-adaptive evolution." Stebbins's book didn't address neo-Darwinism's implications for desert evolution, but he tackled the subject in a 1952 paper:

> The new combined attack . . . has brought into striking relief the great differences which have existed between the rates of evolution of different groups of organisms, or even within the same group in different periods of geological time. On the one hand, we have the types which have remained stagnant for millions of years . . . and on the other hand the examples of very rapid, apparently "explosive" evolution. . . .
>
> Evolution is not a series of even, steady trends, directed from within the evolutionary line, or from without by some preconceived plan or supernatural force. Rather, it is opportunistic, depending upon the interaction between the hereditary variations which happen to exist in the evolving population at any one time, and the environment which happens to surround the population at that time. If we accept this premise, then we can legitimately ask the question: under what conditions of the environment can we expect maximum rates of evolution, and under what conditions will evolution become greatly retarded or even completely stagnant?

Stebbins's answer to this question was the opposite of Darwin's conclusion that evolution occurs mainly in more favorable environments:

"Static populations have nearly all existed in environments which have remained relatively constant and continuously favorable. For land plants, these are chiefly the great forest belts. Extending this reasoning, we might conclude that the most rapid evolution would occur in habitats which are changing, and in particular those which are limiting or deficient in some factor essential to the existence of the organism." Stebbins thought such habitats would include deserts, and he saw three main reasons why evolution would be rapid in them:

> In the first place, where moisture is the limiting factor, local diversity in topography, soil, and other factors will have a much greater effect on the character of vegetation than in regions where moisture is adequate. . . . The second reason . . . is that . . . the most favorable structure for rapid evolution is that of a large or medium sized population divided into many small subunits or colonies which are largely isolated from each other, but can interchange genes through occasional migration between colonies. . . . In arid or semiarid regions, with their regional diversity, this type of population structure may be expected. . . . The third reason . . . is the number of different specialized structures which plants can evolve for adaptation to dry conditions.

Places like the Providence Mountains manifest Stebbins's three reasons. I was struck there by the difference between the sandy plains where it rains least and the rocky hills that get a little more precipitation. The plains vegetation is mainly creosote bush and burroweed with more diverse shrubs along washes; the hills have rock gardens of cactuses and other succulents with occasional pinyon pines or junipers and even ferns and mosses. Each desert range has a characteristic flora, and even the plains vary. The climate provides many chances for incidental migrations—sandstorms and flash floods carry seeds to unexpected places; spiny or sticky seeds cling to animals that range far to find food. Plenty of specialized structures occur; in fact there are few plants without specialized structures. It can indeed look as though desert evolution is furiously riding off in all directions.

Darwin saw the fact that rain forests and coral reefs have more species than deserts as evidence that competition-driven evolution is more active in them. Stebbins saw that fact as evidence that such "favorable" habitats cause *less* evolutionary activity, allowing species to accumulate instead of dying out from competition. Of course, competition would occur in favorable habitats, but in a more limited way. It would take the stress of major environmental shift to accelerate evolution into what neo-Darwinians started calling a quantum rate, sudden bursts of change punctuating a normal slower rate prevalent during periods of environmental stability. The Andes' dramatic rise, which Darwin had associated with the South American desert in 1836, was just such a shift according to Stebbins's view. And so was that of the California mountains.

THIRTEEN

A Neo-Darwinian Galapagos

Nowhere do the California mountains rise more dramatically than in Death Valley. It is the lowest, hottest, driest place in the United States, and it looks it, a vast salt flat walled with escarpments so steep they seem to ascend with visible speed through the heat waves that distort the air. But to me the most dramatic thing about Death Valley isn't the blazing salt but little animals that live, to a degree, in the salt.

My strongest memory of the valley is not of the famous scenery but of happening, after a morning of being roasted and sandblasted by the famous climate, on what appeared to be a mountain stream flowing through a coastal marsh. The patches of pickleweed and salt grass bordering it were white with salt and smelled like the seashore, but the little stream looked as clear and sweet as a Sierra meadow one. Meadow birds flew around—a killdeer, a common snipe, and a black phoebe that wagged its tail happily as it banged a silvery insect against the ground in preparation for eating it.

When I looked into the stream, lively swarms of water beetles and caddis fly larvae on submerged plants enhanced the arcadian impression, as did the darting of tiny fish. Walking along the bank, I was surprised to see that the water literally swarmed with the fish, although the wind-ruffled surface made it hard to get a good look at them. Eventually I found a sheltered spot where I could watch the fish chase each other in what seemed like mating behavior. Most were a mottled honey color, but many were pale blue with yellow heads and black-tipped fins. Some of the blue fish were quite small, but still very active. The honey-colored fish chased each other; the pale blue fish chased each other. The fish's mad pursuits made the stream appear not just lively but frenzied, more like a hormone-driven singles' party at the beach than a Sierran Arcady.

William Manly and his party were among the first white explorers to see similar fish as they made their thirsty way into Death Valley in 1849. They found the discovery encouraging, if not particularly impressive: "One night we had a fair camp, as we were close to the base of the snow butte, and found a hole of clear, or what seemed to be living water. There were a few minnows in it, not much more than an inch long. This was among a big pile of rocks, and around these the oxen found some grass." As the fish implied, the water proved potable, so they passed a rare comfortable night and then moved on.

If seeing the fish in the marshy stream I encountered on the valley floor had encouraged the Manly party to drink, however, they would have been unpleasantly surprised. Whereas the fish in the rocky hole they found inhabited freshwater, the ones I watched in that bucolic-looking stream spend at least parts of their lives in water twice as salty as seawater. Yet all the fish belong to the same genus—*Cyprinodon*—and they are all called desert pupfish.

The fish the Manly party found may have been the Devil's Hole pupfish *(C. diabolis)*, a different species than the ones I saw in the stream, the Salt Creek pupfish *(C. salinus)*. Both species are confined to a single place on the planet. The Salt Creek pupfish inhabits a few miles of that marshy creek. The Devil's Hole pupfish lives on a small limestone shelf at the top of a water-filled fissure in the rocks. The two places are a few dozen miles apart as the crow flies, but the two species look and act quite differently.

Devil's Hole pupfish are shorter and slimmer than Salt Creek ones, with larger heads in proportion to their bodies, and they lack their pelvic fins and banded markings—although I didn't see them up close because their pool is gated to protect them. Over the years, their population can fluctuate between a few hundred and a few dozen. Their small habitat and food supply limit their numbers and they have had to adapt to unusually warm water, averaging about 93 degrees Fahrenheit. The Devil's Hole environment is fairly stable, however (except when humans have interfered), so the pupfish undergo little competition for food or mates.

The Salt Creek pupfish population fluctuates annually from a few thousand for most of the year, when water is scarce, to many thousands when spring runoff fills the creek and adjacent marshes and food is abundant. Competition accelerates as population soars, and sexual activity peaks—as I saw. The pale blue fish were males chasing rival males away; the honey-colored fish were females chasing away females; the smaller blue fish were sexually precocious young males trying to get in on the action. It was a wonder any of them found time or space to actually mate.

These two species are among more than twenty isolated pupfish populations in the Death Valley watershed today. They include at least ten taxa: four species and six subspecies of a fifth species. Each species and subspecies has evolved distinctive forms and biotic adaptations. The Devil's Hole and Salt Creek species are unusual because of the extreme conditions they undergo. Those under less stress behave more "normally" for the genus, with dominant males establishing and defending territories that females visit for breeding.

I watched bright blue males of a more "normal" species—the Amargosa pupfish—guarding their territories in large freshwater springs at Ash Meadows east of Death Valley. They seemed a lot calmer than the Salt Creek males, as though resting complacently between conjugal visits. Surprisingly deep and clear for their bleak surroundings, the springs

were even more full of life than Salt Creek, including another native
fish—a reddish silver minnow called speckled dace—and tadpoles and
crayfish as well as water insects. They also contained illegally dumped
guppies and mollies that have contributed to the extirpation of another
endemic, the Pahrump killifish *(Empetrichthys)*. I could see why Devil's
Hole is gated. A sign said that the springs are the local human water supply as well as the home of vulnerable species, but tourists were smugly
skinny-dipping anyway.

Death Valley's fish became famous as examples of evolution-in-progress
at about the same time Stebbins published his "quantum" accelerated
desert evolution ideas. A 1949 article on the region proclaimed:

> When a biologist stands on the floor of a great Pleistocene lake, where
> strand lines are clearly visible on the mountainsides all around him,
> but whose waters have shrunk until a few short streams and isolated
> springs are the only remaining traces of fresh water in the midst of a
> desert, his imagination is likely to be captured by the occurrence of a
> dense population of fishes and other aquatic organisms in those waters.
> Knowing the importance of geographic isolation and small breeding
> populations in evolution (as deduced from genetic theory), he cannot
> overlook the fact that a natural experiment on the mechanics of specia-
> tion has been performed for him on a colossal scale.

Biologists began seeing these "islands" of aquatic life as a kind of desert Galapagos. In 1979, two fish specialists wrote: "Just as biological
processes and resultant divergent forms of terrestrial life on the remote
Galapagos Islands strongly influenced the thinking of Charles Darwin,
recent studies of the aquatic islands of Death Valley have increased our
knowledge of the evolution of life forms. . . . Evolution and differentiation of pupfishes and killifishes in the Death Valley System has been
influenced by three primary factors: stresses of the environment (selection pressure), length of isolation, and habitat and population size. These
three factors are the forces driving the process of evolution."

Comparing Death Valley to the Galapagos has its limits, however.
Confronted by tiny fish isolated in a salty stream or tiny pool in the
desert, Darwin probably would have seen an evolutionary backwater
where the fish interacted more with the harsh environment than with

other fish, and where natural selection was correspondingly slow. The two biologists' "three primary factors" influencing desert fish evolution sound more like the neo-Darwinian "three main reasons" that Stebbins cited in his 1952 article on accelerated desert plant evolution than Darwin's ideas.

Stebbins's first condition for accelerated desert evolution, scarcity of moisture, is certainly a limiting factor on fish. This scarcity has divided desert pupfish into many isolated populations, thus providing his second condition, a lot of genetic variability. The biologists didn't mention his third condition—specialized structures preadapted for desert stresses. But desert pupfish and killifish clearly have great potential to live in salty, oxygen-poor, hot, or otherwise stressful water.

Devil's Hole pupfish have provided a striking example of this. When biologists moved some of them from their rocky ledge to refuges as part of a conservation effort, new generations of the transplanted fish soon grew longer, with smaller heads in relation to their bodies than their parents. In effect, they began to be more like the "normal" pupfish in streams and larger springs because refuge conditions were more like those habitats, with cooler water and more food. This implied that the Devil's Hole pupfish's ability to survive in hot water with little food is not just an adaptation gradually acquired through natural selection, but a preadaptation, a genetic potential that can be expressed under extreme conditions like Devil's Hole, and then switched back off again if water cools and food becomes more abundant.

Yet if Death Valley's fish supported Stebbins's ideas about desert as an evolutionary frontier, they didn't really address the question of California desert origins. The limits of likening Death Valley to the Galapagos arise again here. Biologists can surmise that Galapagos iguanas evolved relatively recently on those islands (or earlier volcanic islands) because Galapagos iguanas live nowhere else. Death Valley fish are less unique.

The Pahrump killifish *(Empetrichthys)* is (or was) an endemic genus, but close relatives inhabit warm springs in Mexico. Speckled dace *(Rhinichthys)* has congenerics throughout North America, and the pupfish genus, *Cyprinodon,* has species scattered through the southern United States and Mexico and on the East Coast as far north as Maine. The pupfish family, the Cyprinodontidae, also occurs on other continents, and some South American and African genera are adapted to survive complete evaporation of waters they inhabit. They lay their eggs in the mud before dying, and the eggs hatch when the next rains fill the pools again. They're called annuals, like wildflowers.

California desert fish aren't so specialized, which might imply that they haven't been adapting to dryness for as long as others. Or it might not. Salt Creek pupfish survive the winter, when most of their stream stops running, not by laying their eggs in bottom mud, but by sheltering in it themselves. Such things hint that *Cyprinodon* and other fish may have inhabited ancient California deserts as well as present ones. Recent genetic studies suggest that Death Valley's pupfish have been isolated from other North American *Cyprinodon* populations for millions of years.

California desert pupfish generally are thought to descend from ancestors that inhabited a brackish estuary, the Bouse Embayment, in the lower Colorado Valley and Salton Sink regions in the mid-Pliocene epoch. Pre–ice age fish fossils found in Death Valley and other parts of the California desert include *Cyprinodon* and relatives of *Empetrichthys,* as well as other kinds of fish that still live in the Southwest—suckers, squawfish, and Gila chubs. As the Bouse Embayment withdrew southward and the Pliocene climate dried, the fish presumably lived in smaller bodies of water. Some may have been isolated in desert springs and salt creeks.

During the subsequent rainy ice age, however, the ancestors of today's desert fish would have inhabited the then ample lakes and streams, not desert springs and salt flats. Judging from the transplanted Devil's Hole pupfish's quick reversion to pupfish "normality," a transition from isolated Pliocene springs and creeks to huge Pleistocene lakes would not have been a big problem for those ancestors. As their genetic variability implies, pupfish may have been getting in and out of hot water for a very long time. Although they can change so quickly in response to environmental shifts, Death Valley's fish may be as much about endurance as change, perhaps another instance of desert biology going in circles.

FOURTEEN

Mexican Geneses

If pupfish didn't show how California desert originated, that didn't mean other organisms couldn't. Daniel Axelrod was energetically exploring such possibilities when G. Ledyard Stebbins's ideas about accelerated desert evolution emerged. Axelrod suspected that, contrary to what botanists from Asa Gray onward had thought, it was not *today's* desert vegetation that originated in Mexico and moved northward with drying climate. It was another type of vegetation, which he called the Madro-Tertiary Geoflora to complement Asa Gray's temperate-zone Arcto-Tertiary flora of conifers and deciduous hardwoods.

Axelrod thought the Madro-Tertiary Geoflora had evolved from humid tropical and subtropical forest that, as fossils showed, had covered the southern third of North America in the Paleocene epoch just after the dinosaur age. It was essentially woodland, although it included areas of savanna and brush, and it contained the ancestors of the sclerophyllous or "hard-leaved" plants that cover California's central and southern coasts today—oaks, laurels, pines, cypresses, junipers, buckthorns, manzanitas, and sages. Axelrod named it the *Madro*-Tertiary flora because similar vegetation now grows on Mexico's Sierra Madre and includes plants like wild avocado and fig that are extinct in Alta California but that he found in the fossil floras at the Mojave's edge.

Axelrod thought Stebbins's ideas helped explain the evolution of hard-leaved Madro-Tertiary plants from rain-forest ancestors soon after the Paleocene ended fifty-five million years ago. "From an evolutionary standpoint," he wrote in 1958, "the inferred setting for phylads [groups] ancestral to Madro-Tertiary plants would have been particularly well-suited for mega (or quantum) evolution. . . . Rapid evolutionary divergence of ancestral Madro-Tertiary phylads from more mesic tropical types would be promoted by isolation, owing to their restriction to scattered drier sites . . . habitat discontinuity produces moderate-sized or even small populations in which isolated demes [populations] interchange genes only occasionally, a situation favorable to genetic drift and hence rapid evolution."

Madro-Tertiary vegetation had spread north into California's present desert areas in the late Eocene as climate dried. Through the subsequent Oligocene and Miocene epochs, roughly twenty-five million years, its mosaic of woodland, savanna, and brush covered the area, intergrading to the north with Arcto-Tertiary conifers and deciduous hardwoods. In this unimaginably long time, climate often shifted—from tropical, to subtropical, to warm temperate, then back to tropical, then back to subtropical. But it never became really dry because the region remained relatively low lying.

Volcanoes and hills rose and eroded, but they weren't high enough to exclude moist air from the ocean. Even when climate dried even more in the Pliocene epoch that began five million years ago, oak woodland covered the Great Basin, while oaks, chaparral, or an association of small thorny trees like mesquite and acacia grew over what are now the Mojave and Sonoran deserts. This lasted until about three million years ago, when the "climatic accident" of the ice age began: "Late Pliocene and Quaternary elevation of the Sierra-Nevada and Cascade axis and the Peninsular Ranges of Southern California rapidly brought drier climates into existence over the lowlands to the east. Only then did the present regional desert climax come into existence, its species being derived from those represented in the geofloras which earlier had dominated over the lowlands."

California's desert flora, then, was not an ancient invader from the south, but a recent descendant of an ancient invader from the south. It had evolved quickly from the old Madro-Tertiary Geoflora, presumably because of the same three factors—environmental diversity, scattered populations, and specialized structures—that had caused the Madro-Tertiary Geoflora to evolve from rain forest millions of years earlier.

Mesquite and acacia, shrubby trees common in the Mojave and Sonoran deserts today, support Axelrod's thinking. Both are ancient genera that left fossils as far back as the dinosaur age as well as in many California Madro-Tertiary sites. Both occur in diverse environments, often in scattered populations. Both have specialized structures—thorns, drought-resistant leaves, extensive root systems—that preadapt them to desert life. Mesquite can put its roots down dozens of feet into the ground to reach water. It is not hard to believe that honey mesquite and catclaw acacia, the species I saw in the Providence Mountains, could have evolved from nondesert ancestors in a few million years, a blink in geological time.

Mesquite and acacia don't necessarily look like recent descendants of woodland ancestors in places like Mitchell Caverns, where creosote bush and burroweed surround them for hundreds of square miles. Then they can just seem part of earth-old desert. But there are places where a Madro-Tertiary woodland past seems closer. At Big Morongo Canyon, west of Joshua Tree National Park, in the transition zone between desert and coastal vegetation, dense patches of mesquite and acacia overhung with cottonwood, alder, and willow grow adjacent to sparse expanses of creosote bush, *Lycium,* and other desert plants in the mosaic pattern that Axelrod envisioned for the landscape of five million years ago.

Big Morongo did seem like an evolutionary flashback as I sat through the noon heat in a thicket by its stream. Birds I didn't recognize came down to drink along with the usual desert hummingbirds, doves, verdins, flycatchers, finches, and sparrows. Of course, they were simply unfamiliar species. The Big Morongo Canyon Preserve has one of the highest levels of migratory bird diversity in the United States and gets many rarities and vagrants. But, amid the unusual greenery, they might have been relics of a moister, more diverse epoch.

A problem with Axelrod's theory, though, was that unlike mesquite and acacia most living desert plants don't have known fossil ancestors in Madro-Tertiary sites or anywhere else. Jerzy Rzedowski, a botanist affiliated with various institutions in central Mexico, was quick to point this out, along with other problems. Rzedowski seems to have been something of a botanical conservative compared to Axelrod and Stebbins. At least, he concentrated on the tradition, represented by Forrest Shreve and Ivan Johnston, of working to understand living vegetation in situ before delving into the ambiguities of fossils and genes.

Rzedowski had arrived in Mexico in 1946 at age twenty from eastern Poland and Ukraine, where his experiences after 1939 might well have inclined him toward a venerably peaceful tradition like field botany. With its own legacy of atrocity-ridden conflicts, Mexico may not have seemed quite the place for pacific inclinations. Rzedowski's father, who'd arrived with him, soon left again for Israel. But the younger man found Mexicans tolerant, freer of ingrained hatreds than Europeans. And he found plenty of botany to describe and classify. Mexico has one of the world's richest floras, and much of it remained unstudied in midcentury. Rzedowski became a leading expert and decided that, despite the scarcity of desert plant fossils, the extant Mexican vegetation did not support Axelrod's ideas of recent desert origin.

Axelrod's fossil evidence that some American climates had been less arid in the Tertiary did not prove, Rzedowski asserted, that this had been true throughout the period:

> On the contrary, the phytogeographic data seem to incline more toward prolonged existence in Mexico of arid climates necessary to development of a flora as individualized and diversified as the one we have now. Of the 93 apparently endemic living genera in Mexican arid climate zones, 68 are restricted in their distribution to those zones, or have a majority of

species concentrated there, from which one might surmise that many of those genera probably originated in the desert. If to that one adds the fact of the existence in Mexico of genera characteristic of diverse arid regions of the world (*Frankenia* [salt heath, a semishrub of obscure family], *Larrea* [creosote bush], *Menodora* [the olive relative I saw at the Providence Mountains], *Peganum* ["garbancillo," a creosote bush relative], *Thamnosma* [turpentine broom] etc.) one can't sustain the idea of a single recent origin from more mesophilic ancestors.

Rzedowski also challenged Axelrod's claim that Stebbins's theory of accelerated plant evolution in dry environments supported his recent desert paradigm. He pointed out that Stebbins's theory, while convincing in many ways, did not correlate with Axelrod's fossil evidence, which suggested that relatively *little* rapid evolution of woody plants had occurred in the apparently semiarid conditions of the Miocene and Pliocene in the American Southwest. Ten-million-year-old tree and shrub fossils weren't all that different from three-million-year-old ones.

Rzedowski quoted Stebbins's classic *Variation and Evolution in Plants* to that effect: "The floras of the middle and latter part of the Tertiary Period contain only modern genera, and an increasing number of fossils from deposits of these ages are indistinguishable from living forms. In the Miocene Tehachapi flora of southern California, which is an association of largely sclerophyllous shrubs indicating a semi-arid climate, most of the species are likewise closely similar to living ones."

Rzedowski maintained, further, that Stebbins's theory had limited correlation with his own studies of Mexican flora:

> Of the genera studied in the present work, some Cactaceae and Compositae show a geographic distribution which might be seen as the result of rapid evolution in relatively modern times; the majority, nevertheless, and especially the monotypic paleoendemics (*Acanthothamnus, Orthosphenia, Pachycormus* [elephant tree], *Sericodes* [another creosote bush relative], *Simmondsia* [jojoba] etc.) seem to be good examples of bradytely [slow evolution], suggesting they are relics of an ancient desert flora. For Stebbins's hypothesis to be correct, the phenomenon should only be applicable to groups that, by their genetic structure, are predisposed to a rapid external transformation. Accelerated evolution, thus, is not enough to support the idea of a young age for all arid-adapted North American plants.

Rzedowski concluded that of the three main elements of Mexican woody plants—northern, southern, and endemic—the endemic element correlated most with arid climate. "The abundance of the endemic element in the Mexican xerophilous flora indicates that the origin and diversification of this flora took place in ancient times. Such a fact agrees essentially with the conclusions drawn from fossil floras discovered in the southwestern United States, but does not confirm Axelrod's hypothesis, based on the same paleontological data, which postulates a recent (Pliocene-Pleistocene) age of desert climate and of desert plants in North America."

It's unclear what Axelrod thought of Rzedowski's objections to his Madro-Tertiary paradigm since he didn't publish any detailed response to them, although he mentioned them later. They were marginal to his field of paleobotany, and they came out in Spanish in a Mexican journal. Anyway, he was busy responding to a challenge that soon came from closer to home—surprisingly close.

FIFTEEN

Desert Relicts

About halfway along the road from Twentynine Palms to Cottonwood Springs in Joshua Tree National Park, there is an interpretive stop called Ocotillo Patch. It consists, predictably, of a patch of ocotillos, tall woody plants that for most of the year are simply clusters of spiky poles. The patch might seem redundant for a roadside attraction, since ocotillo typifies California's Sonoran Desert and pervades vast areas. Yet the few big specimens at Ocotillo Patch are remarkable because they are the only ones around, as I found one morning when I walked to the nearest foothills looking for more.

The patch lies where the road descends from the Mojave's Joshua tree zone into the Sonoran's Pinto Basin, so I expected to see more ocotillos in that direction. But the only others I saw that day were after I drove six miles south toward Cottonwood Springs and then walked for three hours

up a canyon called Porcupine Wash. Just before I turned back, as startlingly turbulent black clouds abruptly filled half the sky, I found another
ocotillo patch about the same age as the first—mature plants, quite large
for the species. One had fallen over recently, exposing shallow roots.

I don't know why ocotillos are so patchy there. But the patchiness
was striking, particularly since they were undergoing their annual few
weeks of growth. Small triangular leaves along their stems and sprays
of tubular scarlet flowers at their tops made them a nexus for hummingbirds, mockingbirds, and other life that seemed incongruous in the
ambient scrub and evoked a sense of isolation echoing Mary Austin's:
"Go as far as you dare into the heart of a lonely land." They might have
been holdovers from a stranger world than that of the smoke trees and
catclaw acacias that sprawled everywhere along Porcupine Wash. They
seemed like vegetable dinosaurs looming fantastically among botanical
burros and bighorns. A chuckwalla sunning lumpishly nearby might
have been some weird Cretaceous land crocodile.

Smoke trees *(Psorothamnus spinosa)* and catclaw acacias are bizarre
enough by nondesert standards. Smoke trees are gray tangles of thorny
stems that indeed resemble puffs of smoke; catclaw acacia's newly
sprouted leaves are such an improbably garish green that they might
be plastic. But both species are in the huge pea family and are so surrounded by relatives even in the loneliest desert that they evoke a sense
of familiarity.

Ocotillo *(Fouquieria splendens)* is quite alone in the Alta California desert as far as family relations go, although it has a few congeners in Baja like
the "Adam tree" that puzzled Miguel del Barco. The head of the family is
Baja's cirio, Spanish for "candle," a word that aptly describes the plant's
shape and the spray of yellow flowers at its tip. Growing up to seventy-
five feet tall, cirio (either *Idria* or *Fouquieria columnaris* depending on the
authority) is so odd that a botanist nicknamed it the boojum after a mysterious creature featured in Lewis Carroll's nonsense poem, *The Hunting of
the Snark*. Barco considered cirio the ultimate in desert peculiarity:

> Finally, there belongs to the class of pulpy plants a tree which we do not
> know if it exists in any other part of America or the world, nor was there
> any notice of it in California until the year 1751, when Father Fernando

Consag, as result of a trip made to the north between the Sierra and the ocean, published its existence and properties. . . . It grows without any branches whatsoever, and it grows straight and to a great height like the palms, but it does not have (like these) a crown even on its highest reaches. It only has some thin twigs all over its trunk clear to the top . . . these are full of little leaves, each of which ends up in a thorn. At the very top of these trees grow their flowers in a bouquet, but they have no fruit at all, nor can any use be made of them, not in wood for construction, and also not even firewood.

When Consag tried to make a campfire of "that scant and extremely light wood," the smoke had given him a headache.

Nobody knows why the ocotillo family is so lonesome, but one possibility is that it is the last of a formerly much larger group now nearly extinct. This suggests that it may be very old, and although known fossils don't prove this, ocotillos do appear antiquated because their characteristics are so idiosyncratic. They reach tree size but it's hard to see them as trees because they're so skeletal, which makes it just as hard to see them as shrubs. They seem to predate current vegetable categories. There is nothing quite like them, although people often think they are spindly cactuses because their "wood" is similar. As Barco wrote, "They are made of nothing else but a pulp or mass that is supported by a framework of tubes similar to that of the cardon." But their flowers are quite different from cactuses', and they have no surviving relatives away from North American desert and semiarid habitats.

Not surprisingly, ocotillos are among the plants Jerzy Rzedowski cited as evidence that today's California desert evolved in Mexico, perhaps as long ago as the dinosaur age, instead of in California a few million years ago. His critique of Daniel Axelrod's Madro-Tertiary theory, published in 1962, might seem to have complicated the desert origin problem sufficiently. But the problem got more complicated a few years later, when G. Ledyard Stebbins published a paper that also questioned the role of plants like ocotillo in Axelrod's theory.

As Stebbins's recollection of his attempts to get Axelrod into the National Academy of Sciences suggests, the two were friendly, having met at Berkeley in the 1930s. Academic unorthodoxy eventually propelled both to the University of California, Davis, a Sacramento Valley agricultural school that might have seemed rustic for scientists of their stature. But both found it congenial and stayed for the rest of their careers. Stebbins moved there in the 1950s when invited to create a genetics department, and he made it a center for field-oriented evolutionary studies. Axelrod arrived in the 1960s when, according to Stebbins, the geologist with whom he was closest at UCLA "couldn't get along with the other people in the department there, and Dan followed him up."

Axelrod's fossil evidence on the age of California floras impressed Stebbins: "I have discussed all these questions with him very extensively, particularly since he's come back here to Davis but even before then. I have the greatest admiration for him." Although discussion with Axelrod didn't invariably lead to agreement, that did not cause acrimony in this case, perhaps partly because Stebbins was one of the century's more modest, generous scientists.

His own estimation of his best-known achievement suggests the modesty: "So, my book that really put me on the map, *Variation and Evolution in Plants,* did not have any really new ideas that others hadn't talked about, but it put a whole lot of things together in a coherent picture which caused everybody to read it excitedly." I can attest to the generosity. Encouraging comments that he made about a book of mine incited me to present him one of the first copies, a prospect he probably found less exciting than I did. But he met me in his office at the appointed time, although I noticed pajama bottoms under his trousers. He was in his late seventies then and absentminded—he once drove across California with a dead rattlesnake left on his car during a field trip. He must have remembered my visit at the last minute and rushed over so as to not disappoint me.

Stebbins's kindness didn't obscure his critical faculties. In 1965, he and another UC Davis botanist, Jack Major, published an article, "Endemism and Speciation in the California Flora," that addressed themes not unlike Rzedowski's in relation to Axelrod's ideas. It observed that there are two main concentrations of relict plant species in the state: "one in the Siskiyou-Trinity mountain area of northern California, and the other along the northern and western margins of the Colorado [Sonoran] desert, from the Little San Bernardino Mountains along the east slope of the San Jacinto and Santa Rosa Mountains, the Borrego Valley area, and southward into Lower California." The northern relicts include trees like weeping spruce and Sadler's oak; the southern relicts include shrubs like turpentine broom, creosote bush, and ocotillo.

Stebbins and Major agreed with Axelrod that the Siskiyou-Trinity tree species are relicts of the Arcto-Tertiary flora, the forest that once covered most of North America and Eurasia with a greater plant diversity than today's. But they disagreed with his idea that the Sonoran Desert relict shrubs evolved from a Madro-Tertiary flora of oak woodland, savanna, and thorn scrub in the last few million years. They thought them too numerous and desert adapted to be such recent scions of woodland vegetation, even an arid-adapted one. The shrubs seemed more likely to be relicts of ancient, widespread desert floras:

> In the first place, several of the relict species . . . have their closest relatives in the arid regions of southwestern Eurasia and northern Africa, while others, although possessing relatives elsewhere in the southwest, belong to groups which are represented also in the arid and semi-arid regions of the Old World. Because of the close relationship between the Old and New World representatives of these groups, and the fact that the entire genus, or in some cases the family to which they belong, is adapted to semiarid or arid conditions, the possibility that these elements became adapted independently in the two hemispheres to arid conditions, and that their common ancestor belonged to a mesic flora, is very remote. . . .
>
> Secondly, some of the most common species of the area are relict types related to species of the arid areas of temperate South America. . . . In all of these examples, the great similarity of the North American species to those of South America and in some instances also Africa suggests that they have evolved very little since these vicarious elements became separated from each other, and that they existed essentially in their present form when the Madro-Tertiary flora developed in the middle of the Tertiary Period. Furthermore, they are adapted to arid or semiarid conditions wherever they occur, and they have no recognizable connections with any genera of the humid tropics. Consequently, these groups probably also migrated from one continent to another as members of an ancient xeric or semi-xeric floristic element, which preceded the development of the Madro-Tertiary Geoflora.

Stebbins and Major thought it "highly unlikely" that closely related desert plants occurring on two or more continents could have reached such a wide distribution within the time when Axelrod said the Madro-Tertiary Geoflora evolved. They pointed out that most such plants remain confined to warm climates today. If the plants had migrated between the Old and New Worlds, they would have had to do so across areas such as the Bering Strait, which were too cold for them during at least the past thirty million years:

> Another group of genera which are hard to reconcile with the hypothesis that the Madro-Tertiary Geoflora is derived from the humid neotropical flora includes *Lyonothamnus* [ironwood], *Crossosoma* [one of the genera I mentioned earlier that has an entire family to itself], *Simmondsia* [jojoba], and the two genera of the Fouquieriaceae [ocotillo and cirio]. The relationships of these genera are so obscure that they are best explained as remnants of distinctive groups which flourished before the modern families had reached their present state of development. They, also, probably existed in their present form when the Madro-Tertiary flora arose. Their connections with families of the humid neotropical flora are tenuous at best.

In effect, Stebbins and Major suggested that plants like ocotillo don't just seem like vegetable dinosaurs, they *are* vegetable dinosaurs. They maintained that the presence of so many such relics in California desert was "best explained by assuming that their ancestors were for a long time confined to a restricted, homogeneous area, and isolated from their nearest relatives by long stretches of territory unfavorable to them." This led them to a different hypothesis than Axelrod's:

> During the early part of the Tertiary Period and the latter half of the Cretaceous Period when the dominant floras of the middle latitudes in the northern hemisphere were mesic and tropical, subtropical, or warm temperate in character, the existence is postulated of several small pockets of xerophytic or semi-xerophytic floras, similar to the caatingas [areas of scrub or dry forest growing on poor soils] of modern Brazil, and some of the valleys of the Colombian Andes. . . . Such isolated patches of xerophytic species existed not only in southwestern North America, as postulated by Axelrod (1958: fig. 2). They probably extended also from central North America to Alaska, and across Eurasia to the shores of the Tethys Sea, which occupied the present Mediterranean region. . . .

The Madro-Tertiary flora is postulated as having derived through an amalgamation of these isolated elements through spreading from these isolated areas, as aridity increased and conditions for them became more favorable. Some genera had already lost so much of their stored genetic variability that they were unable to react to these more favorable conditions by evolving new species and genera. These became the relict groups. . . . Others were able to resume active speciation, and became the large complex of the Madro-Tertiary flora.

Combined with Rzedowski's challenge, Stebbins and Major's article might seem a crushing rejoinder to ideas that Axelrod had been elaborating for two decades. Like Rzedowski, they brought up an entire group of plants that his confident fossil analyses had ignored. Of course, Axelrod had to ignore them because he found no fossils of them. But it is harder to see how plants like creosote bush and ocotillo could have come from the Madro-Tertiary woodland-savanna vegetation that Axelrod postulated than mesquite and acacia. For all their adaptations to drought, acacia and mesquite are "normal" looking shrubs or small trees that can compete in woodland shade, not bunches of spiky poles.

Still, Stebbins and Major didn't reject Axelrod's work. They didn't challenge his fossil analyses or his idea that the Madro-Tertiary Geoflora evolved in southwestern North America. They merely suggested some adjustments of its geographical and geological elements, as they diplomatically observed at the end:

The hypothesis just presented differs little from that proposed by Axelrod (1958). He did not, however, consider evidence suggesting early migrations of semi-xerophytic warm temperate or subtropical types from the Old World to the new and vice versa. That such migrations took place seems rather likely. Once that assumption is made, then many elements of the Madro-Tertiary flora which are congeneric with elements in the Old World steppe and Mediterranean scrub floras . . . are best explained as having migrated to North America from Eurasia, or vice versa, by means of the 'stepping stones' provided by these small pockets of semi-arid conditions. In view of this possibility, it seems more appropriate to postulate that the Madro-Tertiary Geoflora had a diverse origin, rather than a principal derivation from a humid New World tropics.

In any case, challenges to his ideas did not discourage Axelrod. On the contrary, they stimulated him, judging from the volume of work he would produce in the next four decades. Few scientists have published in quite such bewildering abundance. The pre-1960s publications seem almost meager compared to the later ones.

SIXTEEN

Madro-Tertiary Attitudes

Despite his attention to the details of fossil floras and the strength of his opinions about them, Daniel Axelrod was flexible. A legendary example of his flexibility, and of his prolific energy, was his response to the plate tectonics revolution of the 1950s and 1960s, when new studies of the continents and ocean basins suggested that they are not permanent features, as geologists had thought, but have shifted considerably in the past, driven by obscure forces in the earth's core and mantle.

In 1963, Axelrod published a closely reasoned article that used fossil floras to refute plate tectonics: "The distribution of forests . . . suggests that the continents probably have been stable, not drifting," because "successively younger floras at the same latitudes show similarity in composition, not the changes that would be expected with continents drifting across many degrees of latitude and hence of climate." A decade later, he published a closely reasoned article that used fossil floras to support plate tectonics: "The recent discovery of new facts which demonstrate beyond reasonable doubt that the crust of the earth is mobile, demands that biologists look again at their explanations of some old problems. Manifestly, as climates on land areas change in response to shifting positions of oceanic and continental plates, new opportunities for adaptive radiation may arise."

Axelrod responded with similar flexibility to the Stebbins and Major article on desert relicts. Two years after it came out, he published an article, "Drought, Diastrophism, and Quantum Evolution," that, while not quite saying that he might have mistaken the age of some desert plants, acknowledged that the two other scientists had some good ideas on the subject, ideas that reflected his own: "Evolution of angiosperms in response to drought commenced long before Eocene time because numerous plants representing woodland, chaparral, and thorn forest are recorded in later Eocene and Oligocene floras, and some occur in those of Paleocene and late Cretaceous age (Axelrod, 1950, 1958). Furthermore, the pre-Tertiary evolution of many unique plants that now occur in dry regions may be inferred from their taxonomic relationships."

Axelrod cited as examples of such ancient drought-adapted angiosperms a number of plants Stebbins and Major had mentioned in 1965: isolated genera such as crucifixion thorn (*Holacantha*), jojoba, and ironwood that belong to groups with fossil records going back to the Cretaceous but that are so different from their relatives (crucifixion thorn is related to ailanthus, jojoba to boxwood, ironwood to rose) that they must have diverged long ago; and entire plant families such as cactuses and ocotillos that are so different from all other plants that they must have been adapting to drought even longer.

Axelrod's pro–plate tectonics article, published in 1972, posited new ways in which shifting continents and other features of plate tectonics might help to solve the desert origin problems that Stebbins and Major had raised. They had said that many relict desert plants seemed too isolated and strange to have evolved from the Madro-Tertiary Geoflora in the past five million years. But the evident great age of many desert plant families didn't necessarily mean that the plants had migrated to

California from other continents in the way Stebbins and Major postulated. Axelrod saw no fossil evidence that relict desert plants moved between the Old and New worlds via a northern land bridge when climate was warmer in the early Age of Mammals.

If continents drifted, however—if, for example, Africa and South America had once been closer together—that raised other possibilities. The relict plants might have evolved as part of an ancient arid-adapted flora in the supercontinent of Gondwana that occupied the Southern Hemisphere during the dinosaur age. As seafloor rifting broke Gondwana into Africa and South America over the past hundred million years, and as South America drifted toward North America, some of the ancient desert relicts could have been rafted northward. And the relict plants need not have been part of a vast desert like the present ones. As Stebbins and Major acknowledged, they might have lived in small arid pockets among moister flora. Axelrod thought they even underestimated the possibilities of tectonically generated arid pockets:

> Stebbins (1952) has shown that there are several reasons why plant evolution would be relatively rapid in arid to semiarid regions. . . . While I do not deny the efficacy of topographic relief and climatic change in promoting rapid and divergent evolution, there are nonetheless alternatives to Stebbins's analysis which have not been discussed previously. These are suggested by the evolutionary potential which may be inferred for landscapes of crystalline rock that provided arid and semiarid edaphic sites through Mesozoic and Cenozoic times not only in seasonally dry areas but in wet tropical regions as well.

Axelrod noted that areas of thin soil and infertile granite bedrock support pockets of "desert" plants even in today's humid tropics. Warm Miocene California could have been similar: "The domelands of the southern Sierra Nevada, and those from central San Diego County southward into Baja California, have also probably had a critical role in evolution in that region because they have been exposed since the late Cretaceous. In this area, the trend to greater aridity during the Tertiary has selected plants able to stand increasing drought. It seems highly probable that the new taxa which evolved did so in dry local sites provided by exposed slopes of basement rock."

Axelrod's response to Stebbins and Major's criticism was a reasonable one. "Although he was not always right," wrote a colleague, "he understood clearly how science and the method of multiple working hypotheses worked. He was excited to propose alternative hypotheses and knew from Darwin . . . that it was very valuable to do so and reject those that failed."

Intellectual flexibility can also be construed as opportunism, however, and Axelrod's enthusiasms could seem bossy. Another botanist friend at UC Davis, Michael G. Barbour, recalled, "He would often pull me aside during a hallway or walkway encounter to relate something ecologically new that he'd seen in the field, hoping to convince me, other faculty, or graduate students to drop whatever research we were currently doing in order to follow through on his observations."

Perhaps partly because of such attitudes, resistance to Axelrod's expansive hypotheses from supporters of the "old earth-feature" desert paradigm continued for many years. "Except for a few species (less than 1%)," wrote A. W. Johnson, a botanist at San Diego State University, in 1968, "we are totally lacking in any substantive information on the age, origin, or evolution of the California desert flora. To trace their evolutionary roots to the tropics is reasonable in the sense of the proposed tropical origins of angiosperms per se, but the proximal origin of the majority of species and perhaps genera of desert angiosperms in California apparently lies in the western United States."

Although Axelrod's fossils comprised much of what little was known about California desert plants' evolution, and although Johnson agreed with him that they probably had tropical ancestors and that many had adapted to desert "in the western United States," he doubted Axelrod's theory that they did so recently. Of desert relicts, he wrote:

> Many of these taxa are highly modified vegetatively and, from all indications, have been living in arid habitats for a very long time. It is difficult to rationalize this group of species with the suggestion that their adaptations to deserts are recent. . . . Those few with vicarious populations elsewhere, like *Larrea* [creosote bush] and *Eurotia* [winter fat, a shrubby saltbush relative that also grows in Asia], e.g., are separated from them by thousands of miles. In the absence of clear evidence

of long distance dispersal, their present separation must be a reflection of considerable age. It has been proposed that they originated in small 'arid pockets' at a time previous to that when deserts were widespread in North America and thus were preadapted to extensive deserts when they appeared. Although it is not now possible to deny this possibility, other explanations may be forthcoming as more information on ancient landscapes is contributed from paleogeography and paleoclimatology.

In general, one may conclude that desert climates are very old and that they have existed continuously for a very long time, probably since pre-Tertiary times. However, it [the study of plant fossils from California desert] does not give any useful information as to the recency of deserts as major land features as has been concluded from other studies (e.g. Axelrod, 1950).

SEVENTEEN A Friendly Land

Scientific rivalry clearly played a part in the resistance to Daniel Axelrod's ideas, and his attitude to nonscientists didn't help either. Not only prickly with peers, he disdained to court the popular media. His articles are relentlessly technical, replete with jargon, neologisms, Greco-Latinisms, passive voice, and other barriers to general readership. Unlike G. Ledyard Stebbins and Jerzy Rzedowski, he didn't publish books except for academic monographs. Resistance to his neo-Darwinian concepts of recent and rapid desert evolution had deeper roots as well.

Despite its scientific influence, neo-Darwinism has never engaged the public as "paleo-Darwinism" has. New books, articles, and films about Darwin and his circle emerge regularly: Theodosius Dobzhansky, Ernst Mayr, George Simpson, and Stebbins remain obscure outside scientific circles. After decades of neo-Darwinian teaching in schools, the old

Darwinism of T. H. Huxley and Herbert Spencer retains a much stronger hold on the popular imagination. People can't seem to get it through their heads that natural selection applies to organisms with genes, not, as with Spencer's Social Darwinism, to human cultures and civilizations. Rightists and libertarians still invoke "the survival of the fittest" in politics and economics. Leftists and egalitarians still deplore it. This obdurate linkage of natural selection with culture is one reason why creationists reject the ape ancestry that Darwin proposed in *The Descent of Man* more passionately than their great-great grandparents did.

Neo-Darwinism's synthesis of natural selection with genetics, biochemistry, paleontology, ethology, and ecology is too complex to provoke strong public reaction. Even many highly educated professionals are vague about it. One of the noisier recent struggles arising from old anti-Darwinian passions involved the Dover, Pennsylvania, school board's attempt to mandate the teaching of creationism in the guise of "intelligent design," challenged in court in 2005. The federal judge assigned to rule on the case expressed amazement at the volume of neo-Darwinian evidence that scientific witnesses presented to him.

This vagueness has extended to public attitudes toward desert evolution. The most popular mid-twentieth-century writer on California desert was Edmund C. Jaeger, who assumed his predecessors' role of interpreting it to tourists and campers. After exploring the Mojave with a burro, Jaeger taught biology at Riverside Community College on the desert's edge for over thirty years. He began publishing books on desert in the early 1920s, and his writing influenced creation of the first preserves—Death Valley National Monument and Anza-Borrego Desert State Park in 1933, Joshua Tree National Monument in 1936.

"Our beautiful deserts are wholly the creation of nature," Jaeger wrote, as though tacitly dismissing the earlier writers' mélange of religiosity and Darwinism. "Some of them are among the most appealing of our scenic wonderlands. Their broad basins and long intermountain valleys, their bizarre land forms such as volcanic buttes, mesas, bajadas, and often barren but majestic mountains rising like colorful spires from the low sweep of creosote bush and sagebrush, are places which, left undisturbed, minister greatly to the pleasure and enoblement of man's mind."

Jaeger explicitly dismissed the "murderous brood" version of desert life: "It has become a habit among writers to describe the desert as a region of desolation, cheerless and dreary, a land of relentless heat, with every plant vested in thorns and every animal poisonous and savage. They have dwelt upon the difficulties and perils of travel in mule-and-wagon days. . . . As a matter of fact, the desert is on the whole a friendly land, its beasts no fiercer than those found elsewhere; nor is travel in it, except in rare instances, unusually dangerous for those who use discretion in taking care of themselves and their motor cars."

Jaeger wrote with authority—his knowledge of desert life was both wide and deep. He knew that virtually every desert shrub species in California supports an exclusive species of thrip, a tiny parasitic insect. He knew that some fungus species in California desert also occur in the Sahara. He became famous for finding that a desert bird, the poorwill, can survive winters by hibernating instead of migrating. He was familiar from long experience with just about every species of bush and lizard, and with the relations between them: "The chuckwalla is a thorough-going vegetarian but may occasionally eat insect larvae. During the days of spring it is a greedy feeder on flowers and it even climbs into small shrubs to get them. Several times I have seen it high up among the fat stems of the brittlebush, snapping with heavy jaws at the yellow blossoms. When flowers are not available, both leaves and green stems are eaten. Even such bitter, aromatic shrubs as the burroweed and the creosote bush are not shunned."

Yet Jaeger seemed little interested in desert origins. His books don't ask why so many lizard and shrub species occur in California desert, or why some lizards have taken the unusual saurian course of eating the inedible-looking shrubs. And they don't confront neo-Darwinism as the earlier desert writers had wrestled with Darwinism.

Jaeger's major book, *The California Deserts,* was the most popular guide to the region as it went through four editions from the 1930s to the 1960s. In the 1965 edition, he briefly outlined the fossil record of Paleozoic marine invertebrates and Tertiary vertebrates, and he described in detail the ice age hydrology of desert lakes and rivers. But he didn't mention Stebbins's theories about accelerated or "quantum" desert evolution, Axelrod's ideas on Madro-Tertiary desert origins, or Rzedowski's objections to them. He reprised Asa Gray's 1884 vision of a circumpolar Arcto-Tertiary forest retreating before a Mexican desert invasion:

> It is indeed possible that from Europe and Asia the ancestors of a good proportion of our northern flora was acquired. The botanists tell us that the deserts of North America have derived their plant genera from a very different source. The genera so common in the north are here replaced by such odd xerophilous plants as yuccas, agaves, ocotillos, and cacti, all plants of strictly American distribution that were first developed on the arid Sonoran Plateau of Mexico. It was near the end of the glacial epoch and the beginning of the development of arid conditions in California that these plants of the Mexican realm pushed northward and gave us this most unique flora.

Elsewhere, Jaeger sometimes picked up the thread of an evolutionary question but then dropped it: "In masses of conglomerate rocks found on the desert slope of the San Gabriel Mountains and considered to have been formed in Eocene times are embedded rocks with brown and black coatings of desert varnish showing that arid conditions favorable to the deposition of desert varnish existed even in those ancient times." A shiny veneer caused by chemical reactions between microorganisms and minerals, "varnish" is typical of deserts today, so its apparent presence in California some fifty million years ago would be significant, but Jaeger didn't elaborate.

In another passage, he noted the discrediting of one old anti-Darwinian idea and then fell back on a different version of the same idea:

> Desert plants commonly exhibit certain peculiar structures such as thorns and leathery leaves which are protected against evaporation by hairs, resiny coats, reduced surface, and sunken stomata [the pores through which plants absorb and emit gases]. Such adaptations were long thought to be a reaction to the hot, dry atmosphere and the meager water supply at the root. . . . But this explanation . . . was not adequate. Recent studies carried on by Dr. Kurt Mothes of Germany point to the possibility that both desert and marsh and bog plants produce their similar, peculiar structures because of lack of nitrogen in the soil.

If thorns did not originate, as early botanists like Reverend Henslow and Forrest Shreve had maintained, from aridity's effects on branches, then perhaps they originated from nitrogen deficiency's effects on branches.

Jaeger's apparent indifference to the evolutionary ferment around him is hard to explain. Born in 1887, almost contemporary with Daniel MacDougal and Shreve (although he lived until 1982), he perhaps shared that generation's recoil from Darwinian natural selection toward less mechanical and random forms of evolution. The basic Darwinian emphasis on competition didn't exactly coincide with his idea of desert as a "friendly land." And he may have feared, like Mary Austin, that speculations on the desert's past would distance him from an immediate involvement with it. But he never said so in his books.

Another popular midcentury desert writer did address neo-Darwinism to a degree. Like John Van Dyke, Joseph Wood Krutch was an eastern professor—of literature at Columbia—who brought a sophisticated grasp of contemporary ideas to elegant prose about wild places. Like Van Dyke, he had a sharp eye for gaps in scientific knowledge: "With every passing year it becomes more difficult to understand why or how evolution operates. Fact after new fact proves that the whole process is much more complicated than Darwin imagined."

Noting that a unique black jackrabbit inhabited Espiritu Santo Island off Baja's east coast, Krutch asked: "How does it happen that this striking creature is found in abundance here and nowhere else? . . . The surface of Espiritu Santo is not black. There is no obvious advantage to the black jackrabbit in being different from his relatives a few miles away. It seems a mere caprice." The caprice, he replied, was not in the jackrabbit's apparent flaunting of "survival of the fittest" by being a black creature on a red-rocked desert island, but in its genes:

> One must imagine that when the island was isolated from the peninsula in comparatively recent times a melanistic freak happened to be there . . . and the black individual may likely enough have been a gravid female. . . . If this individual had happened to be on the mainland neither she nor any possibly black offspring would have had a very good chance of survival and would never have founded a prosperous line. But no important predator was isolated along with her. . . . Isolation, together with absence of enemies, removed the need for protective coloration. As a result, the tendency to vary from the norm had free play.

Neo-Darwinian black jackrabbits don't explain desert origins, of course, but Krutch addressed that subject too: "Just how long have our deserts been deserts? And the answer is somewhat surprising. We tend, I think, to assume that they are very old indeed. Somehow, they seem ancient, and they were once thought to be almost as old as the earth itself. But it now appears that this is entirely erroneous and it seems to be generally agreed upon that they are actually quite young or, as the jargon of the experts has it, 'deserts are a recent geomorphic feature.'" Krutch clearly knew Axelrod's work. He quoted the same passage from MacDougal that Axelrod had used to support his 1950 challenge to the ancient desert faction. And he deftly summarized Axelrod's original Madro-Tertiary ideas:

> Though very little plant fossil material has ever been collected in Baja, the evidence afforded by plants, both living and fossil, in the southwestern United States is relatively abundant and striking. There many of the living species most characteristic of the southwestern deserts and confined to them (the cacti, the yuccas, and the euphorbias, for example) have as their closest living ancestors members of the same families in the humid tropics. The best evidence is that, beginning about sixty million years ago, or at the end of the Cretaceous, the climate all over the world grew warmer and plant families which had developed in the tropics spread northward.
>
> Presently, however, the colonists met a challenge. When they first arrived, the climate was still relatively moist and in that respect not disastrously different from the climate of their homeland. The slow rise of the Sierra Nevada Mountains reduced the rainfall over much of what is now a desert area and much later the greater elevation of the San Bernardino and San Jacinto ranges intensified the aridity. One evidence of this is the fact that the fossil plants of the earlier period revealed the broad, soft leaves characteristic of well-watered regions, while the latter present only the small, hard leaves characteristic of plants adapted to arid conditions.

Krutch used the perennial favorite of desert writers as a detailed case in point:

> The cactus originated, either in Mexico or South America, as a normal enough plant with slender stems and orthodox leaves like one member of the family (*Pereskia*) still growing in the West Indies and South America. . . . But from a remote period it has always been distinguished by the anatomical peculiarities of the family—notably by a certain floral

structure and the presence of a unique organ called the areole, from which the spines are produced. Members of the family spread both northward and southward, and somewhere, possibly in what is now one of the American deserts, they developed characteristics which enabled them to survive a long annual dry season. Then, when the slow lifting of mountains . . . turned a seasonally dry climate into a perpetually arid region, they were well on their way to their present ability to tolerate not only seasonal but almost continuous drought.

Yet, despite his bow to the Madro-Tertiary paradigm, Krutch's own conception of desert origins wavered somewhere between Axelrod's and Ivan Johnston's. He seemed unaware of Stebbins's "quantum evolution," holding to the old Darwinian belief that desert organisms struggle for existence more with the environment than each other, and that evolution is thus slower in desert than in more "favorable" environments. He gravitated toward evidence of old, if not "earth-old," desert. He cited a possible prickly pear fossil "embedded in deposits of Eocene age" in southern Utah: "If it is really a prickly pear, then a cactus very like a modern one must have developed during an epoch which, according to the latest evidence, must be placed at least forty million years ago . . . because, so the geologists say, the Great Basin section of Utah became drier and drier as the Eocene was succeeded by the Miocene, and by that epoch it might have been as dry as Death Valley is today."

Contemplating ocotillo's giant relative, the boojum, Krutch envisioned a much older "regional desert" than Axelrod's. He thought boojums had been evolving in Baja since the Cretaceous, when the land had risen from the sea; that their ancestors had come from the south along with other desert plants; and that they probably don't occur elsewhere because they evolved where they live now. Krutch thought the Sonoran Desert might have been "something like what it now is since the Miocene, which means for approximately ten million years." That was seven million years older than Axelrod's "regional desert."

Krutch's willingness to interpret the new desert origin theories for general readers was admirable, and his confusion about them reflects on the theories as well as his interpretation. Still, his version is so confused as to seem incoherent: Sonoran-type desert evolved 10 million years ago from rain forest that spread north from Mexico 60 million years ago, although prickly pear lived in Utah 40 million years ago and boojums have lived in Baja since it arose from the sea 70 million years ago. I imagine him muttering Emerson's remark that "a foolish consistency is the hobgoblin of little minds" while he struggled with such material.

Krutch's confusion about neo-Darwinian theories and Jaeger's indifference to them perhaps stemmed less from their thoughts than their feelings. Jaeger spoke for both writers in his sense of the desert as a "friendly land" and in his response to its ongoing industrialization: "Most unfortunate it is that, perhaps more than any portion of our land heritage, our deserts are thought of by many as the most expendable lands we possess, mere wastelands that should if possible be utilized for gain."

Jaeger saw California's extreme desert not as a savage if gorgeous challenge to civilization but as a tranquil refuge from civilization's destructiveness. For all his willingness to confront neo-Darwinian complexities, Krutch felt the same: "Baja California is a wonderful example of how much bad roads can do for a country. . . . Nature gave Baja nearly all the beauties possible in a warm, dry climate—towering mountains, flowery desert flats. . . . All of this has remained very nearly inviolate because very little of what we call progress has marred it." Krutch saw the desert as a refuge, not only for nature lovers, but for "such creatures as the scorpions, lizards, and spiders, who represent ancient ways of life . . . because life presumably began in hot places." He had more affection for the desert's relictual aspects than for its "progressive" ones like the ability to live without drinking. "As for the animals," he wrote, "some of them drink when they get a chance; and I have caught sight of the commonest of the little lizards darting his tongue two or three times into the water of a small ditch dug to irrigate a cultivated shrub." A subsequent account of the kangaroo rat's feat of metabolically reducing dry seeds to water shifted abruptly to a celebration of the venerable boojum.

When Krutch followed the tradition of invoking a Sphinx-like, oracular desert, it was more Egyptian guardian than Greek trickster: "To those who do listen, the desert speaks of things with an emphasis quite different from that of the shore, the mountains, the valley, or the plains. ·

Whereas they invite action and suggest limitless opportunity, exhaust-less resources, the implications and moods of the desert are something different. For one thing, the desert is conservative, not radical. It is more likely to provoke awe than to invite conquest. . . . It brings man up against his limitations, turns him in upon himself, and suggests values which more indulgent regions minimize."

Although they were evolutionists, Jaeger and Krutch shared a longing for a more humanly meaningful natural world than either the old or the new Darwinism might imply. A venerable, conservative desert may well be a better foundation for this than a recent climatic accident. When both writers were in their heyday, however, neo-Darwinian recentness was about to get some strong support from an unexpected source. Espiritu Santo's black jackrabbits may have revealed nothing about desert evolution, but another small mammal would reveal much. In the process, that mammal would support both Stebbins's and Axelrod's ideas, although not, of course, without controversy.

EIGHTEEN Furry Paleontologists

One day, I took a walk up a canyon in the El Paso Mountains, which run northeast from Red Rock Canyon in the Mojave. The El Pasos have produced as wide a variety of fossils as any part of the California desert— not only the Pliocene mammals that John Merriam found in the Ricardo Formation, but petrified wood later dug from the same formation as well as older animal and plant remains from others. I didn't find any fossils that day, but I did encounter one of California's most influential fossil collectors.

The porous rock of the canyon was eroded into many little caves. Peering into one of these piled with sticks and leaves, I was startled to see a rabbit skull staring back at me, evidently placed there by a desert wood rat *(Neotoma lepida)* that had gathered the sticks to make its den. Wood rats are also called pack rats because of their collecting habits. I made squeaking sounds to try to get the occupant's attention, without much anticipation. I've squeaked outside a lot of dusky-footed wood rat (*N. fuscipes*) houses and been ignored. That species, which inhabits dense coastal vegetation, seldom comes out in daylight.

But desert wood rats are more confiding, perhaps because of their lonelier environment. A plump, big-eared individual popped out of the cave and eyed me, disappeared, and then emerged again for a closer look before it had satisfied its curiosity. Although not as furry as a mountain relative, the bushytailed wood rat (*N. cinerea*), it was a pretty mammal— more like a pika than a Norway rat. I continued up the canyon, and when I later stopped to rest on a pile of boulders, I squeaked again. There was no sign of a den nearby, but another wood rat, smaller than the first, peered at me from another cave. I suppose I could have spent the day wasting curious wood rats' time.

The California desert is literally stuffed with wood rat dens because there are a lot of wood rats and because their dens last a long time. They contain everything that their occupants have collected in their lives, also everything they have excreted. Wood rats get all the water they need by eating green plant material, but they don't metabolize it from dry seeds and concentrate their urine as kangaroo rats do, so, like water drinkers, they urinate a lot. The urine dries into an amberlike substance that binds twigs, leaves, and feces—sometimes bones, feathers, and other oddities—into a deposit called a midden. Because of the urine's chemistry, middens are resistant to decay.

Created by generations of wood rats, old middens can fill caves and are curious objects. I've seen small ones that resembled slabs of peanut brittle, and large ones can be surreal looking. It took white explorers awhile to figure them out. William Manly described a possible encounter with one near Death Valley in 1849:

> So we turned up a cañon leading to a mountain and had a pretty heavy upgrade and a rough bed for a road. Partway up we came to a high cliff and in its face were niches or cavities as large as a barrel or larger, and in some of them we found balls of a glistening substance looking some-what like pieces of variegated candy stuck together. The balls were as large as small pumpkins. It was evidently food of some sort, and we

found it sweet but sickish, and those who were so hungry as to break off one of the balls and divide it among the others, making a good meal of it, were a little troubled with nausea afterwards.

The party suspected they were plundering an Indian cache as they'd done earlier with some squashes, an expedient that bothered the scrupulous Manly. One historian thought Indians might have made the glistening balls by processing reed stems into a kind of syrup. Later observers were convinced that they were from wood rat middens, in which case Manly needn't have worried. The Indians knew about those and probably would have been amused to see the greedy invaders eating dried rat urine.

In the 1870s, Edward Cope seems to have been the first scientist to guess what the curious objects are. Some biologists had thought they were deposited by chuckwallas, which, along with a variety of other creatures, do sometimes live in large wood rat dens. Another den lodger, the desert iguana, is known to supplement its diet with wood rat feces, perhaps as a way of acquiring digestive microbes along with extra nourishment. Herpetologists speculate that this may have helped the lizards to become herbivores, since the microbes break down tough cellulose. But herbivorous lizard excrement somehow doesn't have what it takes to make massive, durable deposits.

Wood rat middens eventually afforded a more edifying use than the forty-niners' impromptu snack. Another desert mystery that emerged from early evolutionary studies was that fossils revealed less about vegetation in the Pleistocene ice age than in the older Tertiary period. Deposits of fossil leaves, fruits, and branches such as Daniel Axelrod dug from Miocene and Pliocene strata are scarcer in later Pleistocene ones. The Pleistocene was shorter than the previous epochs so fossil beds had less time to form, and rapid drying of lakes during arid interglacials would have worked against fossilization of plant deposits. Whatever the reasons, the main known sources of Pleistocene plant remains were pollen deposits in various places and ground-sloth dung found in a few caves. Since wind carries pollen long distances and the sloths wandered widely, neither source was reliable for determining a given location's former vegetation.

Passage through a giant mammal's guts can cause plant identification problems as well. When sloth dung first turned up in the Mojave in the 1930s, Forrest Shreve thought the material in it was from non-desert plants. But Philip Munz, a younger botanist rival of the great plant describer Willis Jepson, identified it as originating from Joshua tree and the saltbush species called desert holly. Other dung deposits, at most around twelve thousand years old, suggested that more trees and fewer desert shrubs had occupied the region in the Pleistocene, but there wasn't enough to be sure.

The situation changed in the 1950s when the U.S. government spread a lot of research money around the desert in promoting its thermonuclear agenda. Two of the scientists who pursued this largesse into the Mojave were Phillip V. Wells, a botanist, and Clive D. Jorgenson, a zoologist. Interested in the Pleistocene vegetation mystery, they intended to visit a sloth cave on a 1961 expedition, but first made a side trip to see if juniper trees, living relics of ice age climate, were still growing on top of a certain mountain. Finding no junipers, they stopped to rest in a canyon: "As they sat and commiserated, Jorgensen looked over his shoulder and something caught his eye about thirty meters away—a dark, shiny mass beneath an overhang. He walked over to it, grabbed a chunk, and broke it open, calling out to Wells: 'You've got to see this! This is where all the junipers are!'"

Three years later, Wells and Jorgensen published an article in *Science* predicting that radiocarbon dating of rat midden plant remains would help to solve the Pleistocene vegetation mystery: "The limited foraging range of the wood rat assures that macrofossils preserved in their ancient middens represent relatively local vegetation. Therefore, the middens probably contain more precise information on local paleoclimate than sediments yielding fossil pollen because of the wide dissemination of many types of pollen, especially that of wind-pollinated conifers. Hence, *Neotoma* middens may have unique value."

In 1966, Wells and Rainer Berger, an anthropologist, published another *Science* article fulfilling the prediction: "Seventeen ancient wood rat middens, ranging in radiocarbon age from 7,400 to 19,500 years and to older than 40,000 years, have been uncovered in the northeastern, north-central, southeastern, and southwestern sections of the Mojave desert. Excellent preservation of macroscopic plant materials (including stems, buds, leaves, fruits and seeds) enables identification of many plant species growing within the limited foraging range of the sedentary wood rat." The material showed that, before around nine thousand years ago, southeast California's vegetation was very different from now. Even though the fossil middens were in present desert areas, most contained remnants of juniper, and some contained pinyon pine, mountain mahogany, snowberry, ceanothus, and other plants that now occur only in higher elevation nondesert in the Mojave. A few even contained maple and ash tree remains. "Prevalence of woodland vegetation at moderately low elevations throughout the Mojave Desert as recently as about 9,000 years ago is apparent," Wells and Berger concluded.

The only place in the Mojave where middens from before nine thousand years ago contained desert plants was Death Valley, and even those—from almost twenty thousand years ago—were plants that grow in moister climates than Death Valley's now. In the Sonoran Desert, far south into Baja, fossil middens also contained quite different plants from today's, including juniper, although some of them were desert plants like sagebrush and Joshua trees that now grow farther north.

The furry paleontologists had provided a striking corroboration of Axelrod's prediction two decades earlier that today's deserts would prove to be a very recent biome, and he was quick to acknowledge this. "That regional climate was not desert, as it is today, is apparent," he exulted. "The regional occurrence of juniper woodland is demonstrated by megafossil remains preserved in ancient woodrat middens at approximately twenty locations in the Mojave and adjacent desert areas."

Wood rat paleontologists seemed made-to-order for Axelrod's version of desert evolution, and not just because of their midden collections. According to the scanty fossil record, the wood rat genus, *Neotoma*, originated from a group of rodents that arrived in North America from Eurasia in the early Miocene epoch and then underwent "a modest radiation in the late Miocene, some 7 million years ago." The wood rats now inhabiting the arid West may have evolved even later, perhaps less than a million years ago.

Wood rats seem exemplary of G. Ledyard Stebbins's theory of accelerated evolution in dry areas. Today two wood rat species, *N. floridana*
and *N. magister,* inhabit the United States east of the Mississippi, while
some twenty species live in the West, Mexico, and Central America. The
California desert region has two arid-adapted species—the desert and
white-throated (*N. albigula*)—while two others—the dusky-footed and
bushytailed—live on its margins. This suggested that western wood
rats began to speciate more rapidly as mountains rose and dryness
increased. In accordance with Stebbins's three conditions for quantum
evolution, habitats would have become more diverse, wood rats would
have been more dispersed, and their ability to build dens and get their
water by eating plants (even dusky-footed wood rats in riparian swamps
can do so) would have preadapted them to desert life.

Wood rats may be so preadapted to desert life that desert species
do better than woodland ones, which suffer high mortality during the
wet winters of the eastern United States and Pacific coast. The eastern
wood rats are rare or endangered in some of their range. Desert wood
rats seem almost invulnerable in their massive dens, which they often
fortify with cactus—also one of their main food items. They seem the
opposite of Darwin's idea that desert organisms mainly struggle with
their environment. Desert-dwelling wood rats may struggle mainly with
other wood rats, since they are solitary and new generations are always
trying to occupy good den sites.

Of course, wood rat middens did not entirely elucidate even the two
million years of Pleistocene evolution. Because of the limitations of radiocarbon dating, fossil middens didn't provide information about vegetation before about fifty thousand years ago. They still didn't show how
or when desert plants had appeared in California. In some cases, they
complicated matters.

Creosote bush is an example. As Axelrod's original 1950 paper on
desert evolution acknowledged, its presence in both South and North
America implies "long distance migration." South America has four
creosote bush species to North America's one, and two of the South
American species live in thorn scrub and dry woodland as well as in
desert. Those species are leafier than desert species. This suggests that

Larrea first evolved from nondesert vegetation in South America and that North America's one desert-adapted species got here much later. But how did it arrive?

When scientists began dating wood rat middens, creosote bush remains appeared only in those of the past seven thousand years. This abrupt appearance did not provide much time for the species to have spread north by the gradual dispersal by which plants usually "migrate." It suggested some more drastic mode of transportation. One migratory possibility, originally raised by Darwin, was that birds might carry seeds on their way from one continent to another during their seasonal passages. Darwin did experiments to show that seeds can cling to birds' feet and remain viable in their guts. Axelrod and Stebbins thought this a possible cause for the apparently sudden arrival of creosote bush and other desert plants with puzzling distributions. But there was no evidence for it.

The mystery cleared up slightly when creosote bush remains over ten thousand years old appeared in an Arizona wood rat midden. The species' arrival in North America had been less abrupt than it seemed, and the genetics of its distribution here support this. Creosote bush grows in the Chihuahuan Desert as well as the Sonoran and Mojave, and its chromosome count changes from east to west. Chihuahuan bushes are mostly diploid; Sonoran ones are mostly tetraploid; Mojave ones mostly hexaploid. This suggests that the species has migrated from east to west with chromosomes multiplying as it colonized new regions, perhaps in response to conditions new to the original population. As Axelrod had noted with California desert wildflowers, plants adapted to very dry conditions tend to be polyploids. "The striking cytogeographic differentiation within the North American *L. tridentata,* together with the complete distinctness of the insect faunas on *Larrea* in North and South America," wrote Phillip Wells and another scientist, "suggest that *Larrea* is of some antiquity on the North American continent."

Axelrod's botanist friend at UC Davis, Michael Barbour, thought creosote bush might have migrated from South America in a semiarid "trans-tropic scrub": "Outlying pockets of *Larrea* in southern Mexico, in Chile, and in Peru intimate a historically wider range for *Larrea.* Minor bits of information sharpen that intimation." Barbour observed that North American creosote bush seedling leaves tend to have three or more leaflets instead of two as with mature plants. Since South American non-desert species have three or more leaflets, this implied that our species' ancestor might have spread north with "trans-tropic scrub" and then adapted to desert. "Stebbins (1952) believes that one of the consequences of adaptation to aridity for plants with compound leaves is a reduction in leaflet number," wrote Barbour.

Axelrod disagreed:

> The meager fossil record does not provide evidence for an arid corridor during the Tertiary. . . . Prior to the Pleistocene, arid areas were much smaller, less severe, and more widely separated than at present. Hence the concept of a "trans-tropic scrub" (Barbour, 1969), from which the present ranges of taxa such as *Larrea* may have developed, is unsupported by any evidence. The idea that relatively long 'jumps' are involved in the establishment of the North-South American desert disjuncts is in agreement with the observations that (1.) they constitute only a small proportion of their respective floras, (2.) the animals associated with them in their disjunct areas are almost entirely different and (3.) they are mainly self-compatible [able to pollinate themselves].

North American wood rat middens have yet to prove Darwin's dispersal theory by disgorging a migratory bird mummy with South American creosote bush seeds in its crop. Given their fifty-thousand-year dating limit, middens don't prove ideas of slower dispersal either. Without further evidence, California creosote bush origins are likely to remain obscure. But despite such limitations, and although their fossil preparation technique is not in the best of taste, wood rats have been surprisingly helpful scientific partners in revealing some of the desert's past.

NINETEEN

Dawn Horses and Dinosaurs

At about the same time that wood rats were revealing the youngest fossil evidence of the region's past, more conventional paleontologists were digging up some of the oldest in the wood rat–filled El Paso Mountains. Lying in the immediate rain shadow of the southern Sierra Nevada, the El Pasos are barren even for the Mojave. Their gray slopes look cadaverous when seen from colorful Red Rock Canyon, with just a faint fuzz of creosote bush among the rocks. So it is odd that they produced the earliest proof that southeast California not only wasn't always desert but once supported lush subtropical forest.

In 1960, Malcolm McKenna, a UC Berkeley paleontology PhD who had recently joined the American Museum of Natural History, announced the discovery there of fossils from soon after the dinosaurs' demise: "The oldest continental Cenozoic fossil mammals, fish, crocodilian, and

chelonian [turtle] remains from California, or for that matter from anywhere in the United States west of Utah, have been collected near the
town of Inyokern from a small number of localities."

Collectors had found turtle and mammal fossils in the El Pasos in
1950, but those specimens had disappeared. McKenna had begun looking in 1954, and, after "extensive prospecting efforts," had found a mammal tooth and jaw fragment in a place called Laudate Canyon. The tooth
was that of a condylarth, a primitive group of hoofed herbivores only
distantly related to living ungulates like horses or deer.

It was a promising start to learning what lived in today's California
desert at the dawn of the Age of Mammals, although it was not precisely
the start. McKenna noted that Daniel Axelrod had explored Laudate
Canyon in the 1940s, following up an 1896 report of two fossil plant
species there. Somehow, he had excavated a number of identifiable fossil
leaves "from the water well behind the cabin of C. E. French, veteran
prospector of the region." I can imagine the energetic Axelrod dangling
from a rope as he dug them out.

Axelrod reported on the plant fossils at a Geological Society of America
conference in 1949: "The Mohave [now Goler] formation . . . from El Paso
Mountains was originally considered Eocene on the basis of two plants,
but this assignment has not generally been accepted. An Eocene age is
clearly demonstrated by a small flora (*Anemia* [a fern], *Annona* [pawpaw],
Myrica [wax myrtle], *Persea* [avocado], *Parathesis* [a relative of burning
bush]) whose species resemble those now in the warm temperate to
subtropical forests of Mexico and Central America." Axelrod's inference
that lush forest had grown where only creosote bush does now was
reasonable, but his El Paso Mountains fossils didn't cause much stir. His
paper on them wasn't published except as an abstract in the Geological
Society's bulletin. The scientific frictions G. Ledyard Stebbins recalled
may have contributed to this. Geologists who said the Goler Formation
was Miocene in age wouldn't have liked hearing it was millions of years
older on the basis of a few fossil plants.

Axelrod seems to have backed away from the hard-to-find El Paso
Mountain fossils for a while after that. But McKenna persisted, grimly,
as he reported in 1960: "Prospecting has been intermittently continued

from 1955 to the present, but identifiable specimens are extremely rare. Approximately one man-week of prospecting per identifiable mammalian specimen has been required. The total available collection from the Goler Formation now consists of five identifiable mammal teeth and a scattering of other vertebrate remains." Still, the fossils convinced McKenna that the Goler Formation was even older than Axelrod thought—not Eocene but Paleocene, the first epoch after the dinosaurs' demise, lasting from about 65 to 55 million years ago.

Some of the fossils were less suggestive of subtropical forest than Axelrod's leaves. The fish bones and some of the turtle bones may have been from marine species. Part of a leg bone "of some reptile adapted to powerful digging" suggested a land habitat, but that was about all. The teeth of multituberculates, small primitive mammals named for knobby cusps on their molars, also indicated little more than land origin. Multituberculates evolved long before marsupials or placental mammals and died out fifty million years ago. They may have eaten plants as rodents do, but nobody is sure. Like rodents, they probably lived in both wet and dry places.

On the other hand, the crocodilian was an alligator, so it could have inhabited swamp forest like living alligators. The condylarths, the primitive ungulates of which McKenna found fossils of two kinds, also may have lived in moist forest. Condylarth teeth lack the high enamel crowns and other chewing specializations of later ungulates, suggesting that they browsed soft vegetation instead of a dry habitat's tough herbs and shrubs.

Primitive mammal fossils turned up again in 1965 from central Baja. William J. Morris, an Occidental College geology professor who excavated them, thought the fossils were as old as the Goler Formation condylarths because they included remains of barylambdids and tillodonts, beasts even more bizarre than the Goler ones. Multituberculates and condylarths may have vaguely resembled rodents and deer; barylambdids and tillodonts resembled nothing living. Barylambdids had small, flattened heads and long, muscular tails—reconstructions suggest miniature duck-billed dinosaurs. Tillodonts grew as large as bears but had teeth like rodents, and may have been burrowers."

The Baja site's other fossil mammal was two-foot-tall, four-toed *Hyracotherium,* the earliest known precursor of horses. Since accounts of horse evolution usually feature a succession from early forest browsers to later grassland grazers, *Hyracotherium* seemed another sign that lush forest covered much of today's California desert region for millions of years after the dinosaurs' extinction.

A year later, in 1966, at a site in northern Baja called El Gallo, Morris found even better proof of long-ago forest: "an abundance of petrified wood." He found the wood with fossils of dinosaurs that had lived a few million years before the late Cretaceous mass extinction ended the Age of Reptiles: "The fauna is the only one from the Pacific margin of the continent where dinosaur materials are abundant." The fossils included many bones of a large duck-billed dinosaur, *Hypacrosaurus,* teeth and vertebra of a large predator related to *Tyrannosaurus,* and back plates of an armored dinosaur resembling *Ankylosaurus.* The trees evidently had matched them in size: "Logs 3.6 to 5 meters long were not uncommon."

"The lithology of the El Gallo Formation is indicative of near shore lagoons and playas, a normal habitat for hadrosaurian [duck-billed] dino-saurs," Morris concluded. "Vegetation was much more profuse than that which exists today, and the better drained areas were thickly wooded."

Overall, the fossils substantiated Axelrod's 1949 picture of early Age of Mammals California as supporting humid, warm temperate and sub-tropical forest. The only possible anomaly was that both McKenna and Morris thought some of their California mammal fossils might be of significantly different species than those of the Rocky Mountains, where more abundant plant fossils show that both dinosaurs and primitive mammals lived in humid places.

In the 1980s, work at the central Baja site by another American Museum of Natural History paleontologist, Michael Novacek, seemed to dispel that anomaly. Deciding that its fauna had lived in the early Eocene epoch, a few million years after the El Paso Mountains one, Novacek concluded that it "could be compared favorably with those of early Eocene age in New Mexico, Colorado, Wyoming, and Montana." It even resembled a newly discovered Eocene fauna from the Arctic: "This showed an extraordinary level of similarity among different habitats. Imagine North

America today with the same kinds of animals and plants stretching from the Pole to the equator . . . a hothouse that nurtured a Garden of Eden—with dense forests, lakes and rivers, and diverse mammals, birds, amphibians and reptiles—that has not reappeared over such a wide swath of the earth's surface anytime since."

Novacek and his team found salamander, alligator lizard, and boa fossils at the Baja site, further suggesting a moist forest habitat. The mammals also had included extinct carnivores called creodonts and primitive artiodactyls, early relatives of deer and cattle. The central Baja site remained more Sphinx-like botanically than Rocky Mountain ones, however. The only plant fossils cited from it were seeds of *Celtis*, hackberry, a tree genus that grows in warm temperate and subtropical forest throughout the world today, and also in California desert.

Dinosaur age fossil wood and ice age wood rat middens made convincing bookends for Axelrod's Eocene to Pliocene Madro-Tertiary woodland theory. But other fossils could be more ambiguous. When petrified wood from the Miocene epoch turned up in southern Baja, for example, it suggested that the area had been forest then. But the wood probably belonged to a tropical tree named *Tapirira* that still lives in semidesert southern Baja, so the fossil didn't prove that much. Other evidence could be even more ambiguous and Axelrod's colleagues kept bringing it up, perhaps motivated in part by his prickly character.

TWENTY

Axelrod Antagonistes

I bumped into Daniel Axelrod's prickles once when I was tempted to write about what Darwin called the "abominable mystery" of flowering plant origins. Fossils show that angiosperms appeared in the late dinosaur age, but nobody is sure exactly why or how. When I consulted G. Ledyard Stebbins, he gently warned that it would be hard "to present the evidence and reasoning in a form that anyone other than a systematic botanist, plant anatomist, or paleobotanist could easily understand." Recently found early angiosperm fossils complicated the problem because they were less primitive than some living flowering plants. "If you can make this 'can of worms' authentic, up to date, and interesting to the non-botanist, it will be a major achievement," he concluded, and suggested five experts I should "feel out."

I wrote them and got two responses. Sherwin Carlquist, a curator at

Rancho Santa Ana Botanic Garden, replied with a letter discouraging me on the grounds that, despite the scarcity of fossil proof, the problem was essentially solved. Angiosperms "originated from a group of now-vanished seed ferns," and other historical theories were simply "bad science." An unsolved problem, he added, as though in compensation, was the origin of the Gnetales, the order that includes ephedra, the desert gymnosperm.

Axelrod, the other responder, offered no such polite diversions. He returned my query letter (addressing it to "Mr. Davis R. Wallace") with a curt disclaimer typed on the margin and underlined in red: "I do not wish to get involved in this problem. At present it has no solution for it is all based on inferences, mostly unsound." I abandoned the subject, but neither my sense that Axelrod was probably right nor his basic courtesy in responding quite eased the sting of his brusque dismissal.

Axelrod's friend, Michael Barbour, said that his "sometimes gruff, blunt demeanor was simply a bluff." Barbour didn't specify what kind of bluff it was, but territorial defense—desert wood rat style—is one possibility. According to Barbour, Axelrod liked to say that Darwin "had it all right to begin with," implying a focus on competition. At Davis, he was known for driving a large white convertible with license plates reading "PROF AX." Despite his dismissal of attempts to solve flowering plant origins, he had published on the problem in the 1950s and perhaps didn't care to see his work appropriated.

A large, prickly cactus den is a magnet to ambitious wood rats, however. In 1976, the year Axelrod retired from his Davis professorship, a number of articles on desert origins appeared, as though their authors had been waiting for the head wood rat to step down before wading into the Madro-Tertiary thorn scrub. A paleobotanist wrote:

> Axelrod's model (1950) of gradual evolution of the arid flora and vegetation in southwestern North America from a Madro-Tertiary geoflora, with the most arid forms and the maximal widespread of arid plant formations occurring only during the Quaternary, does not seem to fit well with the evidence provided by the analysis of the Caribbean flora and vegetation. On the contrary, the ideas of Stebbins and Major (1965) about the existence of small arid pockets along the western mountains

from the late Mesozoic upwards, together with a much more agitated
evolutionary history from that time to the Quaternary, are probably in
better agreement with these data, which account for a heterogeneous
and polychronic origin of these elements.

Three zoologists saw evidence for a desert extending well back toward
the dinosaur age. "Desert adaptation seems to have been occurring over
much of western North America throughout much of the Cenozoic,"
they wrote. They cited a number of fossil species that might be ancestors
of living desert ones.

As Edward Cope and Othniel Marsh showed, the Gila monster,
Heloderma, has a fossil record going back at least to the Oligocene epoch.
An Eocene-epoch lizard named *Parasauromalus* might be related to the
living chuckwalla, *Sauromalus. Paradipsosaurus,* from the early Cenozoic
of Mexico, might be related to the desert iguana, *Dipsosaurus.* Whiptail
lizards *(Aspidoscelis),* the commonest medium-size desert genus (one
climbed on my legs to sunbathe as I was sitting in the Anza-Borrego
Desert on a windy March day), have a North American fossil record
going back to the Miocene epoch. The kangaroo rat, *Dipodomys,* has a
fossil relative, *Eodipodomys,* from the Miocene of Kansas.

The zoologists thought that North American deserts might even be older
than South American ones instead of younger as Ivan Johnston had surmised. They pointed out that although the Monte Desert in Argentina has
ten lizard species belonging to a group called the iguanids, they belong to
only two genera. The Sonoran Desert has sixteen iguanid species belonging to ten "markedly differentiated" genera, including chuckwallas, desert iguanas, leopard lizards *(Gambelia),* zebra-tailed lizards *(Callisaurus),*
horned lizards *(Phrynosoma),* side-blotched lizards *(Uta),* and spiny lizards
(Sceloporus). The Sonoran desert also has more snake species than the
Monte desert. Greater diversity implies a longer period of evolution.

An entomologist who studied grasshoppers on both continents found
a similar pattern:

Several observations suggest that North American grasshoppers have
indeed had more time to adapt to desert conditions than those in South
America. First, in the Sonoran Desert, at least five grasshopper species

feed on *Atriplex* [saltbush] exclusively or regularly, but in the Monte a diligent search in numerous large stands in many parts of the desert failed to reveal any grasshoppers on members of this genus. . . . Second, the degree of feeding specialization on *Larrea* [creosote bush] is more advanced in North America: at least six species belonging to five distinct lineages feed largely or exclusively on this shrub, and at least two of them (*Bootettix* spp.) are completely monophagous [eating only creosote bush]. Thirdly, there is greater resource partitioning among North American *Larrea* inhabiting species: in South America, *Larrea* bushes are inhabited at any one locality by only one species, indicating that the ecological compatibility between the different species has not evolved to the same degree. Fourthly, the proportion of endemic species is greater in North American than in South American deserts.

Ancient-desert advocates cited not only Stebbins and Major in support of their ideas, but Axelrod. "It seems then that selection for xeric adaptation has been going on in the southern continent and, from paleobotanical evidence, in North America as well (Axelrod, 1970, p. 310) for more than 100 million years," wrote one of the zoologists, Frank Blair. "However, major climate changes have occurred in the geographic areas now known as the Monte and the Sonoran desert. The present desert floras of these two areas are combinations of old relicts and of types that have evolved as the continents dried and warmed from the Oligocene onward (Axelrod 1970)."

The 1970 article Blair cited was "Mesozoic Paleogeography and Early Angiosperm History," wherein Axelrod had undertaken the ambitious— even for Axelrod—task of explaining the origins of ancient arid-adapted plants that wound up in today's California desert. The article mentioned on page 310 that "angiosperms were already adapted to dry climate by the early Cretaceous (Neocomian to Aptian), as shown by the small, thick, occasionally revolute leaves in the early Cretaceous floras of California." Blair raised the possibility that there might be evolutionary links between living desert plants and very ancient ones that lived in the rain shadow of the high mountains, the "Nevadan orogeny," that existed in California during the dinosaur age—a possibility that would have complicated the Madro-Tertiary Geoflora theory considerably.

Nobody got the better of even an aging Daniel I. Axelrod, however. A colleague recalled: "After his retirement in 1976, he continued his studies as though nothing had changed, going into his office/lab every day to study and write. He always arrived at 5 A.M. in the morning and worked until 4 P.M. He never slowed down, for he considered his research exciting, and much fun!"

Axelrod had already flanked his critics' use of apparent old desert fossils to attack his new desert theory. In a 1975 article, he had pointed out that early to middle Miocene fossils from northern Florida included arid-adapted animals like kangaroo rats, badgers, tortoises, and many snakes and lizards with Mexican and southwestern affinities. They also included spadefoot toads, a genus so arid adapted that it breeds in temporary pools but that still inhabits Florida and the East Coast as well as western desert. "The tadpoles transform at rates that indicate drying puddles," he wrote of the eastern spadefoot, "implying that its breeding habits have remained basically unchanged since the Miocene, though it has been isolated from arid regions where this habit is obligatory for at least 12–13 million years."

Since no evidence exists that Florida and Alabama have been true "regional desert" within the past hundred million years, Axelrod maintained that all these "desert" animals must have lived in small semi-arid patches among woodlands not unlike the Madro-Tertiary floras in California. In the same article, he noted that pollen of ephedra, the shrubby desert gymnosperm, was "sufficiently abundant in the late middle Eocene Claiborne Formation of Alabama to suggest that it probably was derived from shrubs on local drier sites nearby."

Having covered his flanks, Axelrod mounted a two-pronged counterattack on the old-desert forces in 1979. "Is zonal desert vegetation ancient, or is it a relatively new feature?" he asked in an article about deserts worldwide. "Recent discussion of living biota in the Namib (Zinderen, Bakker, 1976), Australian (Beard, 1977), tropical American (Sarmiento, 1976), Argentinean (Solbrig, 1976), and Sonoran deserts (Blair, Hulse, and Mares, 1976) conclude that desert environments are ancient." But he went on:

> Actually, there is considerable fossil evidence that refutes the idea that regional deserts are ancient. In particular, fossil floras recovered from

the present deserts and their bordering regions show that forest, savanna, woodland, and other vegetation zones occupied these areas for tens of millions of years before the present, and that zonal desert vegetation is in fact quite young (Axelrod, 1950 b). In addition, the occurrence of fossil vertebrate faunas in the present desert regions that include arboreal taxa as well as abundant grazers provides further evidence of their youthfulness. . . . Furthermore, their youthfulness is consistent with evidence which shows that unique minerals of chemical origin produced under moist, non-desert environments have distinguished these areas for most of their history. . . . Furthermore, in some cases, the spread of deserts can be correlated with the recent rise of topographic barriers that isolated them from moisture bearing winds.

With earth-old deserts disposed of on a global scale, Axelrod dealt with his own backyard in a seventy-four-page monograph on Sonoran Desert origins, profusely illustrated with photos, graphs, and diagrams:

Since the lowlands of the present desert region were covered with woodland, thorn forest, and grassland into the Pliocene, the deserts of southwestern North America could not have been extensive, as inferred by Rzedowski (1962), Johnson (1968), Blair (1976), and Otte (1976). . . . Local areas probably did support semideserts by the late Eocene, and they certainly were present in the middle Miocene, as discussed for the Tehachapi flora (Axelrod 1939) and implied also by the Mint Canyon flora (Axelrod 1940: Mss). These floras were situated in local basins on the lee of hills which placed them in rainshadows, in drier climates than those indicated by vegetation of nearby areas to the west or north.

Axelrod likened Alta California's Tertiary vegetation to that of south-central Mexico today, where valleys dotted with cactus, acacia, agave, and yucca punctuate hills wooded with oak, pine, and laurel. As botanists like Jerzy Rzedowski had found, such valleys actually contain as much diversity of "desert" plants as the drier lands to the north. Some interpreted the valleys as relicts of an ancient Mexican desert. But the valleys are not deserts, Axelrod maintained, because they also contain many arid-adapted trees that don't live in true deserts, and because the plants grow tall and thick enough to form thorn scrub or dry woodland.

Axelrod acknowledged that, as entomologists like Daniel Otte had found, North American desert does have an unusual abundance of endemic insects such as creosote bush grasshoppers. But he denied that they had evolved in some earth-old Mexican desert:

> As an alternative, it seems more probable that the insects, as with the plants, also originated in local dry sites, notably in dry topographic or edaphic areas which would also favor the origin of unique taxa, and with later environmental change resulting in their narrow restriction as relict forms in the young regional desert. The evidence overwhelmingly supports the inference that *local* dry sites were the areas where taxa now in desert areas most probably originated (cf. Stebbins 1952, 1974; Stebbins and Major 1965). They were preadapted to spread into the expanding dry regions which finally merged (anastomosed) to form the regional deserts of the present day during the late Quaternary.

Like all battles, Axelrod's struggle with the earth-old faction can be confusing to bystanders. Sometimes it gets hard to tell sides. For example, Frank Blair, the zoologist who used Axelrod's own words in support of ancient deserts, wrote in the same article that South American desert anurans—frogs and toads—are more diverse than North American ones. This implied that anurans have been evolving longer in South American desert than in North American, the opposite of the situation implied by lizards, snakes, and grasshoppers:

> The Monte has the greatest taxonomic diversity, with seven genera versus four for the Sonoran desert. Two of the Monte genera *(Odontophrynus* and *Lepidobatrachus)* are truly desert and subxeric genera, but only one North American genus *(Scaphiophus)* fits that category.

> The greater taxonomic diversity of desert adapted South American anurans may be attributed to the Gondwanaland origin . . . of the anurans and the long history of anuran radiation on the southern continent.

"Paleobotanical evidence suggests that the xeric adaptation may have been occurring in South America prior to the breakup of Gondwanaland in the Cretaceous," Blair concluded, "while the North American deserts seem no older than the middle Pliocene." Not only might North American deserts be younger than South American ones, Blair seemed to imply, they might be almost as young as Axelrod said.

TWENTY-ONE

The Midday Sun

Using modern desert animals to elucidate past desert evolution is con-fusing enough in scientific articles. In the desert, it can be bewildering. Coming suddenly face-to-face with a creosote bush–eating grasshopper one windy day, I had trouble seeing how "normal" evolution had pro-duced it, whatever the continent. It was improbably stout for its meager food plant, colored a snazzy black and chartreuse better suited to a NASCAR rally than a dead-looking shrub in the middle of nowhere. It seemed more like something hatched out in a nuclear test site than a product of natural selection, old or new.

California desert frogs and toads can also bewilder. They may be less diverse than South American ones, but they can be pervasive at the right time of year. I've camped in gaunt canyons where red-spotted toads filled the night with ear-ringing song and tree frogs basked with

surprising nonchalance in full sunlight. The frogs lounged within hopping distance of creek pools and shady crevices, but they stayed in the
sun longer than I could. While my skin turned hot pink, theirs faded to
cool grayish white. Their permeable skin, which absorbs water as well as
evaporating it, is thought to insulate them. Still, I never saw any desert
tree frogs soaking up water in a creek. They have a reputation for avoid-
ing water except to breed or dodge predators.

Despite such complications, there is an evolutionary logic to the way
anurans inhabit deserts. As amphibians, the "lowest" land vertebrates,
still largely breeding in water, they seem marginal desert dwellers, and
they are. Most California desert frogs and toads live near permanent or
reliably intermittent water. Only spadefoot toads depend on puddles
that occur briefly at unpredictable intervals, because they can live dormant in the ground for long periods, awaiting the next big rain. And
some places are too dry even for spadefoots.

The other North American amphibian group, the salamanders, is even
more marginal. Until 1970, they were unknown in California desert.
Then a collector discovered two rare species congeneric to the common
coastal slender salamander *(Batrachoseps)*. They belong to a group that
can breed out of water, but that hasn't helped them much in desert. They
live under rocks in a few isolated canyons where dense shrubs overgrow
spring-fed stream courses. Since they also resemble Mexican and Central
American genera, herpetologists see them as ancient relicts "occupying habitats associated with exposures of these ancient rocks for a long
period, perhaps since early Tertiary times." They seem likely relicts of
Daniel Axelrod's Madro-Tertiary woodlands.

California desert frogs and toads also seem likely relicts of a moister
Madro-Tertiary world. As Axelrod observed, even the most desert-
adapted, the spadefoot toads, live in moister places as well. As Frank
Blair and his colleagues suggested, however, the reptiles—the "next step
up" in vertebrate evolution—are harder to place as Madro-Tertiary relicts, not least in that so many kinds are thriving and abundant in today's
California desert.

Reptiles are popularly regarded as desert "naturals" because they *are*
common in deserts and because, anyway, they belong in such godfor-

saken places along with scorpions, tarantulas, and other creepy-crawlies. From a progress-minded evolutionary viewpoint, however, there is no reason to associate reptiles especially with deserts. Darwin's surprise at the abundance of tortoises and lizards on the dry Galapagos reflected this.

Reptiles are most abundant and diverse not in deserts but in tropical forests (little Costa Rica has 68 lizard species, from six-foot iguanas to cricket-size geckoes, and 127 snake species, from twenty-foot boa constrictors to worm-size blind snakes), and with good reason. Their respiration, circulation, and digestion are less efficient than birds' or mammals', so they compete better where food is abundant and digestible, the climate warm and moist. Although they do have nonpermeable skins and amniotic eggs, with embryos enclosed in a membrane that protects them from water loss, they remain more vulnerable to extremes than animals with livelier metabolisms. Deserts are among the planet's most extreme environments, with sparse, coarse food and huge climatic fluctuations.

Then why do more kinds of reptiles live in California's deserts than in its warm temperate coastal woodlands, where food and climate are more equable? Many more amphibians live in coastal woodland than in desert. I've found dozens of salamanders of several genera in a few miles' walk there. If desert reptiles are relics of the Madro-Tertiary flora and fauna as desert amphibians seem to be, shouldn't reptiles be at least as diverse and abundant in California's coastal woodlands?

Some reptiles are abundant in coastal woodland. Pacific pond turtles can throng waterways. Garter and gopher snakes are common sights, as are rattlesnakes. Ring-necked snakes, sharp-tailed snakes, king snakes, and rubber boas make regular appearances. Coastal woodland fence lizards, alligator lizards, and skinks are even commoner than snakes. Still, as I've said, reptiles seem peripheral in coastal woodland, central in desert.

The centrality of some reptiles in the scorching, barren desert makes a kind of sense. Desert tortoises carry insulated houses on their backs, store water in their bladders, and are strong burrowers. If rains fail they simply stay underground, and they live a long time. Their aplomb when encountered is impressive—usually just a perfunctory hiss and casual withdrawal of legs and head—and they were abundant before pet collectors and ORVs. The western Mojave around Red Rock Canyon had as many as two thousand per square mile, and they remain common on the nearby Desert Tortoise Preserve.

Snakes are elusive in desert. A typical encounter was when, after days at Joshua Tree National Park during which I saw not the smallest, drabbest serpent, I almost ran over a big coachwhip near the entrance to the I-10 freeway. Named for their braidlike scalation, coachwhips hunt in the open and so are more likely than most species to at least be glimpsed. I wanted to take that one back into the park, but when I'd swerved and screeched to a stop it was already long out of sight. Harry Greene, a snake expert, marveled that coachwhips "almost defy the laws of physics sometimes" while he tried to catch "an enormous magenta" one that "seemed to fly, not always touching earth."

Apparent scarcity is an aspect of desert snake success, however. Field guides show that California's desert snakes are more diverse than coastal ones, and they are probably more abundant. Alta California's desert has five rattlesnake species for example—its coast has three—two of which are confined to the extreme south and also inhabit desert. Snakes are the most recently evolved reptiles, and their ability to live sinuously and virtually invisibly in the interstices of things is an aspect of their novelty. It gives them a big advantage in a land of heat, drought, and vigilant predators.

Desert lizard prominence is harder to understand. Although most extant lizard groups appeared no earlier than most living bird and mammal ones, they still seem the most primitive reptiles, the closest in body plan to the first amniotic egg-laying sprawlers that evolved from Paleozoic tetrapods. Edward Cope, an early analyst of the "amphibian to reptile" transition, thought the first reptiles resembled "the farm fence lizards of today."

Lizard survival gear seems rudimentary for the exacting desert environment. Most lizard teeth are simple pegs or spikes in contrast to mammals' elaborate tool kit for slicing, piercing, shearing, and grinding. Some like the Gila monster supplement their teeth with poison, but they lack snakes' highly specialized fangs. And Gila monsters' poison doesn't avail them in the superbarren California desert: like giant cactuses, they largely stop at the Arizona border. Herbivorous lizards may have cusped or serrated teeth, but even they process food relatively poorly. Reptile jaws are weak compared to mammals' because they are made of a jigsaw puzzle of small bones instead of a few large ones, so they can't chew very well.

Lizard physiology is surprisingly vulnerable to desert climate, as Raymond B. Cowles demonstrated—callously—in the 1930s:

> I spent a number of days in testing the upper temperature limits of snakes and lizards by exposing them one at a time to the impact of the sun in their natural habitat. The test required my squatting in daytime heat and allowing the tethered animals to run toward but not reach nearby shade. Watching these supposedly heat-demanding animals quickly die from high temperatures in their chosen home climate was a thought-provoking experience. Here was I, a heat-generating animal with a naked, unprotected skin, surviving even longer exposure while dozens of reptiles were killed in minutes by overheating. Some of the smallest reptiles died within about sixty seconds after being scooped out of their underground shelters into the blazing sun of the surface.

Lizards are thus even more dependent than birds and mammals on behavioral expedients to prevent overheating. Smaller ones bury themselves or move into crevices in midday, while larger ones stay near shade. When zebra-tailed and leopard lizards startled me with their midday athletics at Mitchell Caverns, it was because I was startling them from under sheltering bushes. Their goal in fleeing so briskly was to get under another bush.

Nevertheless, like mad dogs and Englishmen, lizards do go out in the midday sun and, like Englishmen, thrive. To quote another nonsense verse from the mad Englishman who coined the word "boojum":

> "You are old," said the youth, "and your jaws are too weak,
> For anything tougher than suet;
> Yet you finished the goose, with the bones and the beak—
> Pray, how did you manage to do it?"

Like Lewis Carroll's goose-eating Father William, lizards have surprising talents. Although their reptilian background prevents them from politely masticating their food, their jaws are well adapted to gaping, grasping, and gulping. And they have methods of dealing with the desert climate besides hiding under bushes. When they bask in the sun during cool mornings, the blood supply to skin and extremities increases, which quickly warms the rest of the body. When they start to overheat, surface blood flow decreases. They can also change overall skin color to stay cool or get warm—becoming lighter in hot places and darker in cool ones. They have a sense organ for such feats on their heads—a small round "pineal eye" that includes a lens, pupil, and iris. It can't really see, but it evidently senses ambient light, because its removal impairs activity and reproduction cycles.

Because they *don't* need to maintain a constant body temperature, moreover, lizards can tolerate a greater range of it than birds and mammals, which helps them to spend cool weather dormant in the ground and hot weather going out in the midday sun. Desert iguanas can amble around eating creosote bush flowers with body temperatures between 104 and 107 degrees Fahrenheit. Raymond Cowles observed that some desert lizard species could survive "an occasional and very brief 113 degrees F," a body temperature he would have found the reverse of thought provoking if he had tried it himself.

Indeed, herbivorous lizards compensate for poor chewing ability by using high body temperature to aid digestion. Chuckwallas normally start feeding at body temperatures of 100 degrees Fahrenheit or higher. A little sunbathing sprawled on a boulder soon raises their ungainly-looking bodies above the desert's much lower nocturnal ambient temperature. Their genus name, *Sauromalus,* translates from Greek as "flat lizard." An eighteen-inch-long individual may have a four-inch-wide abdomen.

Perhaps most important from an evolutionary viewpoint, lizards need much less food than birds and mammals, since their metabolisms use less energy. Cowles certainly would have found food for thought if he'd tried living on creosote bush and burroweed. Lizards can fast longer, and large ones can stay dormant in the ground during droughts, living off their fat. When conditions are right, a given area can support many more lizards than mammals or birds, allowing greater scope for survival and adaptation.

If Darwin had paid more attention to lizards, he might have changed his mind about deserts as evolutionary backwaters. It may be generally true, as he wrote of desert dwellers in *The Descent of Man,* that "nearly all the smaller quadrupeds, reptiles, and birds depend for safety on their colours." But he had described showy orange, yellow, and red coloration in his land iguana specimens from the "very sterile" Galapagos, although he passed it over in his later evolutionary thinking. Desert lizards can wear surprisingly brilliant colors, as John Van Dyke noted, albeit vaguely: "The lizards are many in variety, and their colors are often very beautiful in grays, yellows, reds, blues, and indigoes."

There has been nearly as much argument about desert lizard colors as about cactus thorns. Van Dyke thought they could change to match their surroundings, "having traced with surprise the faintly changing skin of the horned toad produced by the reflection of different colors held near him." This raised the question as to whether they simply react automatically to environmental stimuli—as with the lightening or darkening of skin in response to heat or cold—or can change color in response to visual perception of their surroundings. And then there was the social question. Some naturalists maintained that lizards use colors simply to impress rivals in territorial disputes. Others thought they also use colors to impress prospective mates, precisely the competitive sexual selection that Darwin thought stressful desert life would minimize.

Whatever the reasons, California desert lizards can produce a definite if subtle rainbow effect, especially in mating season. Tiger whiptails are checkered blue-black and yellow, with a strange violet bloom. Spiny lizards have blue bellies and throats along with blue flanks. Zebra-tailed lizards sport pink or orange throats and yellow flanks as well as the blue and black flank bars I saw at Cinder Cone Lava Beds. Desert iguanas show pink and orange along their flanks. Male leopard lizards have pink bellies; females may have orange-spotted necks and flanks in spring, a blush indicating that they already have mated and further suitors need not apply.

The peacocks of desert lizards are collared lizards *(Crotaphytus),* leopard lizard relatives of rocky areas. Named for their black and white necks, collared lizards have big heads that—with their habit of running on their hind legs—make them resemble tiny tyrannosaurs. If dinosaurs were as colorful, they would have been impressive indeed. Local populations within their large range may have green, blue, red, or yellow throats along with blue flanks. As with leopard lizards, gravid females may have orange markings. I've seen peacock blue collared lizards, although California individuals have been less spectacular, with green-yellow speckled bodies and bluish heads. One of these, a juvenile, was earnestly trying to catch a side-blotched lizard not much smaller than itself.

The more scientists observe lizard coloration, the more complex it seems. When Barry Sinervo, a biologist at University of California, Santa Cruz, studied the abundant little side-blotched lizard for two decades, he found a bewildering variety of color "morphs." Males can have throats and flanks colored dark blue, yellow, or bright orange. They can also have striped blue and yellow throats and flanks, blue and orange striped throats with yellow flanks, or blue and yellow striped throats with orange flanks. Females can have throats and flanks colored yellow or bright orange, or they can have blue and orange striped throats with orange flanks, or blue- and yellow striped throats and orange flanks.

Male coloration coincides with mating behavior. Blue males establish long-term bonds with females and stay near them. Yellow males sneak around trying to mate on the sly, and they mimic females to further their philandering. Orange males rush about trying to expel other males and to force their attentions on females. Chivalrous blue males cooperate to guard females, which helps them to exclude the sneaky yellows. Boorish orange males can break up blue associations, but they compete with each other so much that the yellows sneak in and mate with the females.

Female coloration also coincides with reproduction. Yellow females produce small clutches of large hatchlings, which makes their offspring more competitive when local population density is high. Like pickpockets, sneaky yellow males do better in crowds. Orange females produce large clutches of small hatchlings, which makes their offspring more competitive when local population density is low. Like Old West outlaws, boorish orange males do better in scattered populations.

One of Sinervo's associates, Lesley Lancaster, found that the gray and tan markings on side-blotched lizards' backs play a part in all this. Some lizards have sideways bars on their backs, which helps to camouflage them in vegetation. Barred yellow males can sneak around better. Some have longitudinal stripes on their backs, which help them to dodge predators. Striped orange males can swagger about in the open more safely.

Apparently the quantity of a hormone, oestradiol, in embryos affects marking patterns. Experiments suggest that male lizards' colors can influence the amount of oestradiol that females secrete into their eggs, and thus the back markings of the hatchlings. If yellow males predominate near a female, she may produce more hatchlings with barred backs, giving yellow male hatchlings a survival advantage. If orange ones do, she may produce more hatchlings with striped backs. This would have pleased Edward Cope since it implies that, as he thought, organisms can respond to their environment in major physiological ways and pass on the responses to their offspring

Side-blotched lizards' color-coded behavior is like a Mozart libretto, full of sartorial finery and amorous intrigues, disguises and cross-dressing. Hormone shifts can turn yellow males into blue ones—tenors into baritones. And how do the multicolored lizards get along? Do males with blue and yellow striped throats and orange flanks vacillate between gallantry, sneakiness, and boorishness? Do similarly spangled females produce similarly confused offspring? If side-blotched lizards try to practice Darwin's basic sexual selection by choosing the most attractive mates, they must have complicated love lives.

I can't say I've watched side-blotched lizards enough to verify all this. I haven't seen blue lizards behaving particularly chivalrously or yellow ones acting any more sneakily than small lizards usually do when grabby giants loom over them. I can say that I've encountered an orange one that was, if not boorish, at least unusually bold—or foolish. Perched on a rock at the mouth of Porcupine Wash, he or she seemed unintimidated at my approach and almost let me touch his or her striped back before discretion got the better of hormones. But the little lizards' variety of hues must have reasons.

Other, less-studied lizard species probably have unplumbed depths of mating behavior. A theater critic as well as a professor, Joseph Wood Krutch reviewed a zebra-tailed lizard pair's performance enthusiastically:

> Besides the advances and retreats which are the essential features of all courtships, this one consisted principally of poetic speeches or amorous arias, though I could not be sure which since the sounds were completely inaudible to me, at least through the window. The male would mount some two-inch elevation, raise himself high on his front legs, inflate his throat until he looked like a small iguana, and then give voice to some sort of utterance which shook his whole body from head to tail. His lady would listen intently, move a little closer, and then edge away again when her suitor approached to ask what effect his eloquence had produced.

Krutch's zebra-tailed serenade is scientifically dubious. Some lizards can squawk or bark, especially if disturbed. I've heard that the Cahuillas named the chuckwalla for a sound it makes. Only the small, mostly nocturnal geckos are known to make or respond to less rudimentary vocalizations as part of mating. But who knows? "Ears comprise the reception system and are well developed in most lizards," write two lizard specialists. "Other acoustic signals might include seismic signals generated when large lizards interact aggressively: the possibility is virtually unstudied."

Lester Rowntree, a prominent California botanist, was singing to herself while collecting plants one day when a lizard emerged from under a boulder, climbed on her knee, and "showed an enormous capacity for large doses of song, closing his eyes in absurd abandon and opening them whenever I shut up, his eyelids sliding back to reveal pleading orbs. This went on for some time till I finally . . . placed him, limp with emotion, on the boulder."

TWENTY-TWO

Lacertilian Ambiguities

Joseph Wood Krutch wasn't sure that his zebra-tailed lizards' performance was sexual, much less musical. But he did cite indirect evidence of the former: "The performance went on for some two hours. . . . I hope the ultimate result was some more little lizards to continue one of the most ancient lines. That such a happy event sometimes does occur I know, because I discovered, just about the time of this courtship, several inch-and-a-half-long individuals of the same species scampering about in the dry herbage by my patio wall."

When I walked up Porcupine Wash, small zebra-tailed lizards scampered out of my way every few minutes—each curling its tail daintily over its back. Whatever its behavioral components, lizard mating clearly works. Its success in populating deserts actually agrees quite well with a basic aspect of natural selection that even Darwin forgot sometimes.

Natural selection is not directed toward progressive goals like increased intelligence, greater energy efficiency, or even improved adaptation. It may cause such progress incidentally. But the only thing that natural selection directly works toward is successful reproduction. If lizards can reproduce as well as, or better than, birds and mammals in desert, it doesn't matter if they are "ugly . . . stupid . . . lazy."

But did California desert lizards become numerous and diverse in an "earth-old" desert or in a "recent regional" desert? By neo-Darwinian standards, as with pupfish and wood rats, their diversity could suggest quick radiation when geologically sudden aridity dispersed and isolated populations, rewarding "preadaptations" like an ability to be out in the midday sun. On the other hand, a slowly accumulated diversity in keeping with "paleo-Darwinian" notions of desert as an evolutionary backwater might be a simpler explanation. Unfortunately, there is little concrete evidence—as opposed to speculative argument—for either alternative.

Living lizards provide some possible evidence, but it is ambiguous. As Daniel Axelrod and his critics noted, California's desert lizards are largely different from those on other continents. His critics saw this as evidence of ancient separation between the continents' deserts. Axelrod saw it as evidence of recent evolution. Desert bushes have more intercontinental similarities than desert lizards, suggesting that the lizards evolved after the bushes.

There are some links between California desert lizards and others. The very common whiptails belong to a group called the teiids that mainly live in South America today. Strangely, teiids were abundant in North America during the dinosaur age but then became extinct here for most of the Age of Mammals. They probably spread back north when the Central American land bridge began to form during the Miocene epoch. Whiptails are the only living teiid genus that has reached the United States so far, and they may have accomplished this feat because, like side-blotched lizards, they have ingenious reproductive tricks.

Whiptails are among the few vertebrates that regularly practice parthenogenesis, the phenomenon wherein a species dispenses with sexual reproduction. Many whiptail species consist entirely of females that produce eggs without benefit of males, all the eggs hatching out as more females. The reproductive advantage of this is obvious, because the species consists entirely of egg layers that solely produce more egg layers and so on. "Unisexual" parthenogenetic species apparently result from hybridization between sexual species, and since around two dozen known whiptail species inhabit the southwestern United States, there are a lot of opportunities for this. About a third of the known species are parthenogenetic.

All this is very interesting, but its implications for California desert are confusing. California has no parthenogenetic whiptail species. In fact, it has only two whiptail species—the tiger whiptail (*Aspidoscelis tigris*), which occurs throughout California desert, and the orange-throated whiptail (*A. hyperthyra*), which lives mainly in Baja. Their ranges overlap but they haven't hybridized to produce a parthenogenetic species. These facts might be evidence that whiptails have inhabited California for less time than their more diverse congeners in New Mexico and Colorado and have had less time to speciate and hybridize here. It is hard to see evidence of California desert's age in this, however. Both California species live in a variety of habitats as well as in desert. So whiptails could have spread into California when it was an old desert, or they could have done so when it was woodland and savanna and then adapted to a recent desert.

Other links between living California desert lizards and lizards elsewhere are equally tenuous. The chuckwalla's genes and anatomy suggest a relationship to the green iguanas and ground iguanas common from Mexico to South America. A tropical forest chuckwalla ancestor might have adapted first to drier conditions in parts of the Madro-Tertiary flora and then to a "regional desert." On the other hand, despite its common name, the desert iguana's genes and anatomy show that it is only distantly related to green iguanas, ground iguanas, or chuckwallas, so those traits don't imply a tropical forest ancestor. It might have been evolving in desert from the start.

The fossil record is another possible source of evidence, and is also ambiguous. Desert lizard preservation has the usual constraints. Delicate bones don't last long in an environment where scavengers crunch them up and wind knocks them across rock and sand. I've seen tortoise shells in the desert but not lizard bones. The only dead lacertilian I recall seeing aside from road and ORV kills was a zebra-tailed lizard on the floor of the visitor kiosk at the Desert Tortoise Preserve, and its blood was still wet. A kangaroo rat was hopping away with a baby rat in her mouth, so some true-life drama must have transpired, but I don't know what.

As I've said, most fossil lizards found by early paleontologists in the West aren't clearly related to living species. They do include one surviving genus, *Heloderma*, the Gila monster, along with the possible relatives of chuckwallas and desert iguanas— *Parasauromalus* and *Paradipsosaurus*. More recently, a landfill site in the Simi Valley north of Los Angeles yielded lizard fossils from the middle Eocene epoch about fifty million years ago. They include bones of *Parasauromalus,* the possible chuckwalla relative, and of another genus called *Paleoxantusia,* which may be related to little night lizards *(Xantusia)* that live in rock crevices and Joshua tree bark today. The Simi Valley fossils also include remains of primitive arboreal primates resembling today's lemurs and tarsiers, corroborating the other fossil evidence that tropical forest then grew on California's coast. And renewed digging in the Paleocene Goler Formation in the El Paso Mountains turned up fossils of even earlier primates, as well as fossil wood and three lizard genera. But these very old lizard fossils' relationship to living desert species remains tenuous.

Later California fossil faunas do contain definite links to living desert lizards. One of the few places from which Daniel Axelrod didn't publish fossil collections is the Anza-Borrego region that tormented explorers from Pedro Font to Joseph Chase. Among its torments are some of the West Coast's biggest, weirdest badlands, with fossils dating from the late Miocene to early Pleistocene. Chase made his comment about seeing "Mother Earth scalped, flayed, and stripped to the skeleton" after toiling through one for several days:

> Every quarter mile brought some novelty to sight. In crossing a bench
> of reddish clay I noticed numbers of bullets of some heavy metallic
> stuff, the size of marbles and perfectly round. Then came a tract covered
> with pebbles, various in color, but as even in shape and dimension as
> if carefully sorted. Again, plates of clear gypsum, as large as small

window-panes and nearly an inch thick projected from the sides of the gully. Next, stumbling on some lumps of brittle material, I found that they were compact clots of oyster shells (we were a few hundred feet above sea level). It was a region to charm the geologist, though not the botanist. A few wretched creosotes and ocotillos alone held on to life, shriveled, leafless, and half ossified.

I've never had so strong a sense of descending into a desert past as when I dropped down the startlingly steep escarpment that separates Anza-Borrego from the Coast Range summit. Once there, I could walk for miles across rolling, cocoa-colored clay reminiscent of the Great Plains badlands that Edward Cope and Othniel Marsh explored. There was a similar sense of spectral prehistory, as when I once climbed into foothills and glimpsed a pair of desert bighorns on the horizon. The same color as the clay, the elusive sheep seemed as much emanations from the rocks as living beasts.

"During the Pliocene epoch," wrote two paleontologists in 2006,

> petrified wood suggests the Anza-Borrego region was not a barren desert landscape, but a lowland of meandering rivers and riparian streams associated with the ancestral Colorado River drainage. Mixed woodland flora of temperate western hardwoods, namely bay laurels, walnuts, avocados, cottonwoods, ashes, buckeyes, and palms occupied the setting. In this landscape lived camels, horses, and pronghorns as well as deer and peccaries, suggesting that, like the Mojave sites, it was a shifting mixture of woodland and savanna. Kangaroo rats and wood rats inhabited the landscape, but they were more closely related to species that live in moister landscapes today than to Anza-Borrego's present desert species.

Mark Norell, a paleontologist at the American Museum of Natural History, described the reptile fossils: "The Anza Borrego lizard assemblage is taxonomically the most diverse Neogene [Miocene and Pliocene] assemblage yet documented." Twelve fossil lizard taxa are known, twice as many as from any other site. There are five snake taxa and five turtle taxa. But the reptiles' contribution to reconstructing the prehistoric landscape is not simple.

Whereas the Pliocene fossil plants and mammals in Anza-Borrego are largely different from the ones there today, most of the fossil reptiles are like those in today's desert. They include a desert iguana, a leopard lizard, two species of horned lizards, two species of spiny lizards, a side-blotched lizard, and a whiptail lizard, also a desert tortoise, a night snake, a California king snake, a coachwhip snake, and a rattlesnake.

There are differences. A species related to today's tropical green iguana, Novacek's small iguana, lived in Pliocene Anza-Borrego, as did garter snakes, giant tortoises, and aquatic turtles. The fossils also suggest changes in the landscape over the roughly 2 million years of their deposition. Bones of desert iguana and Novacek's small iguana occur in the lower strata, before about 2.5 million years ago, while bones of lizards that inhabit coastal woodland today, such as alligator lizards and skinks, occur in upper strata. This implies that climate became moister during that time. Still, it does not obviate the possibility that lizards now living in desert were living in nondesert some 3 million years ago.

And that was not the only time "desert" reptiles apparently lived in nondesert. Wood rats collect reptile bones along with everything else, and the bones they collected in Alta California's present Sonoran Desert from 14,000 to 9,000 years ago when it was largely pinyon-juniper woodland are surprising. They are mainly from the same reptiles that live with ocotillo and creosote bush in Anza-Borrego today—chuckwallas and leopard lizards, rosy boas and shovel-nosed snakes.

"It was shown that the composition of the herpetofauna has been relatively stable for the past 14,000 years, while the vegetation has changed remarkably," concluded J. Alan Holman, a herpetologist. "Another outgrowth of this study was that biological scenarios postulating that United States desert herpetological species were restricted to Mexican refugia during glacial periods were negated by the presence of desert species during the late Wisconsinan [the last continental glaciation]. . . . This has important implications. Many of us reasoned that herpetological species, being ectothermic ['cold-blooded'], would be much more sensitive to environmental changes than birds and mammals, and thus we often tried to estimate paleoclimates on the basis of herpetological species. This concept certainly requires rethinking and refinement."

If desert iguanas lived in oak woodland in the Pliocene and chuckwallas lived in pine woodland in the Pleistocene, this might challenge the supposition of Frank Blair and colleagues in 1976 that their ancestors evolved in ancient desert. Mark Norell, who studied the Anza-Borrego fossil reptiles in 1989, inferred something of the sort: "The analysis of these specimens indicates that many taxa occupying the area today have a long history in the region. . . . This agrees with the other findings that generic radiation of many of the region's extant genera occurred relatively early and may be associated with the origin of the Madro-Tertiary geoflora in the early Miocene."

Still, evidence that California desert lizards had relatives and possibly ancestors in ancient woodlands doesn't prove that the lizards lived there before they lived in desert. They might have lived in an even more ancient desert before moving into the ancient woodlands. As if to demonstrate this confusing possibility, a number of living desert lizard taxa apparently have moved into woodland quite recently.

TWENTY-THREE

Xerothermic Invasions

Round Valley Regional Preserve east of Mount Diablo in the San Francisco Bay Area is a prime example of what John Fremont extolled as California's "fresh and verdant" coastal valleys. A gentle bowl of savanna and grassland tucked into the edge of the Coast Range, it contains just the kind of riparian woodland I vainly sought at the Kern River Preserve in 1983—a stately gallery of big valley oaks, sycamores, buckeyes, and walnuts as well as cottonwoods and willows. There's even an old fig tree, clearly left over from a homestead but now seeming integral to the preserve's secluded splendor.

I was watching red-legged frog tadpoles in a creek pool one late summer day when a small lizard appeared among the oak roots on a sunny ledge. Assuming that it was one of the usual Coast Range fence lizards, I barely glanced at it, but it kept darting around in a way that began to seem un-fence-lizard-like. I looked closer and saw that it lacked a fence lizard's spiny scales. It was one of the side-blotched lizards about which Barry Sinervo and his coworkers have discovered so many surprising things. Though typical of California's deserts, the species also inhabits the eastern Coast Range north to the Bay Area. Sinervo got his experimental specimens there, which makes one wonder just what color-coding tricks the species may get up to in its much larger desert range.

Mount Diablo has at least three other lizard taxa found mainly in desert—tiger whiptails, horned lizards, and sagebrush lizards, little fence lizard relatives most common in the Great Basin. I've encountered whiptails there several times, but it's always strange to see the same checker-patterned lizards that hotfoot it over the desert sand wallowing coolly in the coastal grass. Many other "desert" animals inhabit drier parts of coastal California. Before farming displaced them, they thrived in the San Joaquin and Sacramento valleys, and they persist in the eastern Coast Range's rain shadow. Some provide bigger surprises than lizards.

During the 1980s I lived in another Round Valley two hundred miles north of Mount Diablo. It is at fifteen hundred feet elevation in the eastern Coast Range, where winter rain and occasional snow support oak savanna on the valley floor and hardwood-conifer forest in the hills. To the east is the Middle Fork of the Eel River—a major salmon and steelhead stream before logging and ranching degraded it, and still habitat for trout, mergansers, otters, pond turtles, and yellow-legged frogs. Walking through the live oaks, blue oaks, and ghost pines along the Eel one damp winter day, I saw a sizeable long-tailed bird fly out of a patch of brush and glide across the river. I at first assumed it was a grouse, but when it landed and darted off into the trees, I could tell by its slimness and the way it raised a crest of head feathers that it was a roadrunner. I saw roadrunners along the river several times after that, recognizing them also by the blue skin patch behind the eye.

Although Round Valley summer temperatures regularly exceed 90 degrees Fahrenheit, I didn't expect to see roadrunners in riverside woods hundreds of miles from any desert. Joseph Wood Krutch described the species as "perhaps the most remarkable of all desert birds . . . which refuses to live anywhere *except* in the desert. . . . Other creatures, including many other birds, elude and compromise. They cling to the mountains or to the cottonwood filled marshes, especially in the hot weather, or they go somewhere else. . . . The roadrunner, on the other hand, stays here all the time and he prefers the areas where he is hottest and driest." This still is an accepted view. According to Round Valley local lore, the roadrunners there were the descendants of a pair someone had brought "from Barstow" in the Mojave.

Yet roadrunners frequent the Sacramento Valley margins a few dozen miles east of Round Valley, and I think the ones I saw by the Eel River were from that population, not Barstow imports. Krutch exaggerated the species' desert preference. Part of a mostly tropical radiation of ground-dwelling cuckoos, California's species also inhabits brush and savanna east to Oklahoma and south to Peru. "Formerly the range of the road runner included the grassy plains, chaparral covered hills, and arid mesas from Kansas to the Pacific Ocean," Edmund Jaeger wrote in the 1920s. "With the settlement of the land and the increase in the number of gunmen, this unique bird is rapidly becoming rare."

Roadrunners were not the only "desert" animals to turn up in Round Valley. Each summer, my fifty-year-old shack on the valley floor produced windscorpions, also called sun spiders. These are yellowish arachnids that look more like crickets than spiders or scorpions except for their large, toothed jaws, which they use to crack open the exoskeletons of arthropod prey. Windscorpions are abundant in African and Asian deserts as well as American ones, suggesting long evolution in that habitat. They only appeared in Round Valley at the hottest time of year, as though they'd "beamed up" from the distant Mojave to enjoy the seasonal ambience. I also saw scorpions around the valley, and tarantulas. Kangaroo rats and horned lizards occur in northwest California, although I haven't seen them there.

Many "desert" plants inhabit the Central Valley and Coast Ranges. Desert olive, rabbitbrush, goldenbush, and scale broom grow in Mount Diablo's rain shadow. Much of the southern San Joaquin Valley was a near desert of saltbush scrub before farming. Some of this remains, inhabited by kit foxes and leopard lizards as well as roadrunners and side-blotched lizards. I've seen leopard lizards running dinosaur-fashion through Coast Range bushes. In the 1970s, G. Ledyard Stebbins and a colleague described another virtually desert vegetation in the Panoche Hills, a part of the south Coast Range's east side that gets so little rain, between six and seven inches a year, that livestock often graze it down to bare soil: "A unique type of vegetation occurs on Panoche Mountain. The only location known to the authors in the world where a non-coniferous gymnosperm dominates a vegetation type is on Panoche Mountain. Dense stands of *Ephedra californica* occur at higher elevations, forming a very chaparral-like community. *Juniperus californica* and *Happlopapus linearfolia* [goldenbush, a sunflower relative] make up the rest of the perennial vegetation. In years of good winter rains, many species of annual plants bloom on the hillsides of Panoche Mountain, with a good number of desert species occurring here."

Stebbins also observed in his Panoche Mountain description that the south Coast Range's rain shadow contains a high concentration of Madro-Tertiary relict endemics. This is reminiscent of Daniel Axelrod's idea that "desert" organisms originally evolved in small dry areas scattered through the Madro-Tertiary flora's woodlands. Places like the Panoche Hills might be "time capsules," small remnants of the piecemeal process that, in the past thirty million years, produced organisms preadapted to recent "regional deserts." The Round Valley roadrunners might be the ancestors of Barstow roadrunners instead of their recent descendants, although they are the same species. The Central Valley and Coast Range leopard lizard—the blunt-nosed leopard lizard—is a different species from the desert one. So is the coastal horned lizard. The coastal tiger whiptail is a different subspecies than the desert one.

Axelrod did not see it that way, however. He thought Coast Range roadrunners and leopard lizards were relicts not of an ancient Madro-Tertiary world but of a recent desert invasion:

A significant, readily recognizable warming trend followed the last glaciation. . . . The intermediate period of warming, termed the Xerothermic period lasted from about 8,500 to 3,000 years ago. Its wide occurrence over temperate latitudes suggests that it was a worldwide event. Evidence for the Xerothermic period comes from two sources. First is the fossil record, most of which is provided by pollen studies of plant succession in post-Wisconsin bogs, which record continuous changes in forest composition in areas adjacent to the sites of accumulation. The second is based on inferences of climatic change which may be drawn from present disjunct distributions of more arid plant communities in areas of moister, cooler climate.

Axelrod thought organisms from the regional desert had migrated coastward as climate dried during the Xerothermic, citing as one example the Joshua trees that I saw in Kern Canyon in 1983: "Coville (1893) noted the occurrence of a large group of desert plants in the dry, upper Kern Canyon, isolated from the desert by the surrounding woodland and chaparral. . . . Climate in the recent past evidently was sufficiently dry to enable them to migrate across Walker Pass and descend into the Kern River Valley to the west, where they have persisted."

If coastal "desert" organisms are relics of a dry period that ended three thousand years ago, it would be conformable with Axelrod's assumption that coastal mountains have risen in the past three million years. Presumably, the big change from the relatively low-lying, subtropical Miocene terrain to the cooler, more mountainous ice ages would have erased earlier landscapes and faunas.

According to Axelrod, then, the Coast Range's "desert" organisms are descendants of "recent regional desert" ones, and differing species like the blunt-nosed leopard lizard have evolved in rapid neo-Darwinian fashion from desert species. Of course, if this is so, then the blunt-nosed leopard lizard and the other "desert" lizards now inhabiting the Coast Range give even less insight into desert lizard origins than the "desert" lizards that apparently inhabited nondesert in Pliocene-epoch Anza-Borrego.

Still, if neither fossils nor most extant lizards show how desert lizards evolved, the situation is not completely hopeless. There is a living genus that seems to provide somewhat better insight into at least one desert lizard's possible origins, albeit ambiguously.

TWENTY-FOUR Sand Swimmers

An area that I found as unexpectedly enchanting as Red Rock Canyon during early desert trips was Algodones Dunes between Anza-Borrego and the Colorado River. Dunes hadn't interested me particularly as natural places; I'd associated them mainly with movies about Rommel's desert war. But another writing project sent me to the Algodones in 1985. When I got there one late afternoon in March, I might as well have arrived at a desert war. One side of the road was solid with parked pickups and trailers, and the dunes beyond swarmed with ORVs from which came occasional gunfire and a steady screeching, grinding roar.

Luckily for me, the Bureau of Land Management had recently closed the dunes on the road's other side to ORVs. So the roar faded as I fled across the sands, apparently the only pedestrian within miles. Although this must have been an illusion, I saw little sign that the machines had been grinding over the dunes for decades before the closure. I guess sand is a good ecological cosmetic. They seemed improbably pristine, and very alive.

First I crossed an orchardlike expanse of big creosote bushes and ephedras—the biggest I'd seen, almost tree size—then increasingly bare waves of rippled sand, pale orange in the late afternoon light, leading up to sharp dune crests. But calling the sand bare is a misnomer. It was written over with hundreds of animal tracks—uppercase swirls of reptiles, lowercase jottings of mammals and insects—and illuminated with wildflowers, their colors enhanced by the orange background—white of big evening primroses and desert lilies, pink of sand verbenas, bluish pale gray-green of their foliage.

The creosote bushes were blossoming and were full of native bees the same pale yellow as the flowers. I noticed little pits in the dunes and saw a bee fly into one. In a moment it emerged and pushed a pinch of sand out, evidently digging a nest. Seeing yet more pits as the wind blew clouds of creosote bush petals overhead, I realized the dunes were full of the nests, full of the creosote bush pollen that the bees were storing.

I saw a movement in a small depression, and a tarantula slowly crawled out and made a peculiar pulsing movement with its abdomen, then dropped out of sight so quickly it might have dematerialized. I began to feel that I'd wandered into the metaphysical essence of desert.

Then I saw something that bolstered the impression. The sand had blown conical mounds around the shrubs, and as I approached one of these, a big sand-colored lizard suddenly materialized under my feet, dashed toward the mound, and vanished into it. As I walked through a group of mounds, this happened enough times that I surmised that I was not seeing vagrant whiptails, desert iguanas, or other generalists. I had found that paragon of California sand dune adaptation—the fringe-toed lizard. A number of fringe-toed lizard species of the genus *Uma* live in North American deserts—three in California—and they are famous for adaptations that allow them virtually to swim in sand as they flee enemies or pursue insect prey. They rate a lengthy description in a global compendium, *Living Reptiles of the World:*

In the first place, the body is flattened, but unlike that of the ungainly horned toad, still elongate, enabling the lizard to knife its way through the sand yet leaving it enough length to employ "swimming" undulations. The tip of the snout is sharpened to an edge by having the lower jaw shorter than the upper and actually countersunk into the upper. The push given by the hind legs in sand swimming is multiplied by the development of a fringe of scales on the toes. As the feet are drawn forward the fringing scales collapse; as the feet are kicked back the scales stick out, thus increasing the surface area brought to bear against the sand. The fringes act like snowshoes when *Uma* runs on the loose surface of the dunes. Finally, the nostrils and ear openings are protected from sand grains by valves and the eyes by fringed eyelids.

Fringe-toed lizards belong to a group that includes some of the desert's commonest species, such as zebra-tailed lizards. Called sand lizards, the group generally has adaptations to sandy soils, such as protected ear openings and burrowing ability, although they live in many habitats. California fringe-toed lizards, however, inhabit only the relatively small parts of the desert where dunes prevail. The only others I've seen were in Death Valley dunes. This might seem to simplify the origin of at least one California lizard genus. If they live only in desert dunes, it would seem that fringe-toed lizards probably evolved in them. And there is circumstantial evidence that they evolved on dunes here relatively recently. They are less adapted to dune life than many Old World dune species. According to three herpetologists: "The specializations that allow *Uma* to inhabit Aeolian sand deposits are behavioral and morphological; physiologically *Uma* does not differ from other iguanid lizards." This seems another evidence of Daniel Axelrod's "recent regional desert."

The *Uma* origin question is not quite that simple, however. When Edward Cope first noticed the fringe scales on its toes in 1866, the genus was known only from California and Arizona, a situation that prevailed until the 1940s. This seemed to support its recent evolution here. Then an *Uma* species turned up in Mexico's Chihuahuan Desert, isolated from the Arizona and California species and different from them.

"The Chihuahuan Desert cradles at its center a huge, hot dry lowland system referred to as the *Bolson de Mapimi*," wrote a herpetologist, David J. Morafka, in 1977.

> Climatic conditions in the Mapimian region may have been highly insulated and continuous since middle Neogene [late Miocene] time . . . virtually all Chihuahuan endemic herpetofaunal groups are either centered in, or restricted to, the Mapimian subprovince by their distributions. Moreover, a striking majority of these endemics are extremely primitive species, often with mixed character states indicating a nearly transitional condition between a mesic ancestral genus and a more arid adapted genus or species group. . . . The example of *Uma* is particularly favorable to the model of Mexican plateau origins, since the Chihuahuan species is more primitive than Mohavian and Peninsular congenitors.

One primitive aspect of the Mexican *Uma*, Morafka noted, is that it is less specialized for dunes than its congeners—even though today's Chihuahuan Desert has dunes. It doesn't burrow in sand as readily or often as the California species. It doesn't just live in dunes: "The eastern or Chihuahuan desert is virtually without a highly adapted dune herpetofauna. . . . Even when sister species are represented in both western and eastern deserts, only the western member is clearly committed to sand dune life. This is true for the sand lizards." The Chihuahuan fringe-toed lizard might provide evidence that *Uma* is older than the highly specialized California species; that the genus first evolved in an ancient Mexican desert and later spread northwest.

But lacertilian ambiguities, like Chinese boxes, seem always to open into further ambiguities. Morafka did *not* interpret the Chihuahuan fringe-toed lizards' primitiveness as evidence of ancient Mexican desert origin. He saw it as evidence of ancient Madro-Tertiary flora origin: "Vegetation in the [Mapimian] region is strikingly similar in both organization and, to a lesser extent, content, to the Mojavia Madro-Tertiary Geoflora from which AXELROD (1958) suggested that a modern desert vegetation evolved. . . . Thus, the Mexican Plateau would be favored as *one* center of origin for Madro-Tertiary vegetation and Young Northern faunas, but certainly not *the* center. The lizard genus *Uma* is a good example where the primitive member is Chihuahuan."

Morafka saw the California fringe-toed lizards' dune specializations as an aspect of Axelrod's recent regional desert:

The underlying cause is linked with the reverse ice age histories of the two dune systems. Refugial western desert dunes actually expanded as glacial sea levels dropped by as much as 130 meters along the coastlines of the Gulf of California. The isolated eastern desert faunas were trapped in isolated valleys or bolsons of the Mexican plateau. During glacial times, pluvial lakes filled the valleys, inundating most exposed dunes, and leaving as refugial habitats only the outcrops, talus, and alluvium that became lake shores. The same climatic events severed the two desert systems as cooling climates and downward-placed vegetation intruded across the Continental Divide, expanded the coastal dune systems of western desert refugia, and simultaneously contracted or obliterated those of the east.

So, according to Morafka, the dune-specialist *Uma* species evolved in recent regional desert, but the genus *Uma* itself evolved in the Madro-Tertiary flora's older semideserts. California fringe-toed lizards—like desert bushes and wood rats—thus seemed to fit the neo-Darwinian scenario on which Axelrod based his ideas. They evidently evolved rapidly in response to regional desert's recent expansion because rising mountains and increasing dryness isolated them and because they were preadapted to spreading dunes.

TWENTY-FIVE

Axelrod Ascendant

David Morafka's interpretation of the fringe-toed lizard's ambiguous distribution in favor of Axelrod's ideas indicated a paradigm shift. Despite
the resistance of biologists like Frank Blair, and perhaps also because of
the confusions involved in that resistance, the Madro-Tertiary Geoflora
had come into its own by the late 1970s. In a 1978 monograph coauthored
with Peter Raven, director of the Missouri Botanical Garden, Axelrod presented his California desert origin scenario more as fact than hypothesis:
"A number of genera that are recorded in late Cretaceous to Eocene floras,
where they are in tropical savannas to thorn forest vegetation, may well
have been present over the area of southeastern California by the Eocene."

> Madro-Tertiary vegetation spread with expanding dry climate follow-
> ing the Eocene . . . and had already assumed dominance over interior
> southern California by the Miocene. The very rich Tehachapi and Mint
> Canyon floras clearly demonstrate that the present Mohave-Sonoran

desert regions formed a single floristic province at the time, one dominated by sclerophyllous vegetation and thorn forest. . . .

During the middle Tertiary [the Oligocene and Miocene epochs], it is
apparent that distinct species of numerous genera were originating, in
Abutilon, Agave, Atriplex, Bursera, Dalea, Euphorbia, Ipomeia, Krameria, Lycium,
and *Yucca*. It was chiefly during the fluctuating glacial-interglacial ages
that numerous species of many genera, especially herbaceous ones, originated in the region. In addition, many mesic taxa that range far out of the
desert province today may well have entered it during the last pluvial,
and were "stranded" in locally favorable habitats as the climate became
progressively more arid. Thus the richness of the desert flora owes chiefly
not to the antiquity of the desert on a regional scale, but rather to the accumulation of numerous taxa during the Tertiary and Quaternary, taxa that
were preadapted to increasing drought over the region.

Axelrod's collaboration with Raven was itself a sign of ascendancy. A
G. Ledyard Stebbins protégé who got a master's at Berkeley and a doctorate at UCLA in the 1950s, then taught at Stanford, Raven became one
of the country's most influential botanists after joining the prestigious
Missouri garden in 1971. He described Axelrod as "the first person who
matched modern plant communities with ancient fossilized ones . . . a
fantastic contribution to our knowledge of vegetation change over the
past forty million years."

By the 1980s, in marked contrast to its former advocacy of ancient
desert, the general paleobotanical literature was endorsing the Madro-
Tertiary paradigm. A 1986 pro-Axelrod article by Robert F. Thorne, a
curator at the Rancho Santa Ana Botanic Garden in southern California,
was a 180-degree turn from the anti-Axelrod piece that San Diego State
botanist A. W. Johnson had published in 1968:

> The present California desert flora is largely composed of autochthonous
> [native] elements from the Madro-Tertiary Geoflora that dominated
> southern California and adjacent areas throughout the Tertiary. Other,
> rarer floristic elements, also preadapted to long periods of drought, hot,
> dry summers, and cool wet winters, have arrived from other parts of
> America and the rest of the world to complete the desert flora. . . .
>
> Many elements have reached the Southwest by relatively normal
> overland movements of propagules [seeds, etc.] over long periods of

time, propelled no doubt by changing climate. Others may have crossed narrowed straits, seas, or even oceans when sea-level was lowered in glacial epochs or much earlier when continents were closer to one another, although later to be forced widely apart by sea-floor spreading and other plate tectonic movements (Thorne, 1978). Changes in oceanic-current patterns surely were sometimes involved. In other instances, as especially in the amphi-tropical American disjunctions, the only logical explanation has to be long distance carriage by migratory birds.

Thorne also saw the desert as geologically young: "If our present California deserts are largely of Late Pliocene-Pleistocene age, the desert plant communities are even more recent, having evolved from existing Madro-Tertiary plant formations as the desert-forming rain shadows developed. These plant communities are of Holocene age, i.e. they acquired their present distribution and composition during the last 8,000–11,000 years since the retreat of the last, Wisconsinan, episode of the numerous cycles . . . of the Pleistocene continental glaciation."

Thorne's article cited seventeen Axelrod papers dating from 1937 to 1980. It didn't cite any by Forrest Shreve, Ivan Johnston, A. W. Johnson, Jerzy Rzedowski, or other ancient-desert advocates. Thorne didn't even mention the idea of ancient desert except briefly: "Those of us without much geological training, in view of our brief life span, tend to think of large-scale features of the landscape as rather permanent, thus very ancient. Yet geologists assure us that large lakes, islands, mountains, and the deserts behind them are relatively ephemeral in the geological time scale. Our southwestern deserts are no exception."

As Thorne's article showed, the idea that today's West Coast mountains reached their present elevation only within the past few million years remained a major support of Axelrod's recent regional desert paradigm. A geology textbook published in 1976 and revised in 1990 continued to maintain that the high-altitude "Nevadan orogeny" covering the Sierra Nevada's present area during the dinosaur age had become a lowland by the early Age of Mammals, and had remained so until a few million years ago: "The erosion that stripped 11 kilometers (7 miles) or more of cover seems to have been largely completed by the close of the Eocene, leaving slowly moving streams draining an eastern crest. The overlying sequences of volcanoclastic and volcanic flows were deposited upon the roots of the Nevadan mountains. Subsequent elevation by faulting rejuvenated the region and produced today's Sierra Nevada."

In 1993, *Assembling California*, the fourth volume of John McPhee's popular account of North American geology, *Annals of the Former World*, echoed this conclusion in the light of plate tectonics:

After the batholith [the massive intrusion of molten granitic material that caused the Nevadan orogeny] came nothing during the many millions of years of the Great Sierra Nevada Unconformity. At any rate, nothing from those years was left for us to see. The rock record jumps from the batholith to the andesite flows of recent time. . . . A few million years ago, when lands to the east of us began to stretch apart and break into blocks, producing the province of the Basin and Range, the Sierra Nevada was the westernmost block to rise, lifting within itself the folds and faults of the Mesozoic dockings, the roots of the mountains that had long since disappeared.

Another indication of Madro-Tertiary ascendancy was that Axelrod himself largely stopped writing about desert origins after answering the opposition with his 1979 articles. A 1983 piece on the subject had a kind of finality:

This brief summary review of paleobotanical evidence shows that regional (zonal) desert environments are not 'earth-old features' as some would have us believe. There is not one shred of evidence to support such a notion. On the contrary, sequences of progressively younger fossil floras from areas presently desert . . . clearly show that these areas were earlier covered with rich forests and woodlands, and that desert vegetation gradually came into existence as progressively drier climate spread during the late Cenozoic. Never in the history of the angiosperm phylum have desert taxa been more abundant, more widespread, or more diverse than at present.

Axelrod kept producing meticulously researched papers, but most were on local fossil floras in areas outside southern California. It was as though he'd finally tired of the secretive Sphinx—early fossils of cactus, turpentine broom, ocotillo, and the like remaining stubbornly elusive—and returned to the simpler, more concrete joys of identifying and analyzing the fossils—mostly of Arcto-Tertiary trees—that he could find.

TWENTY-SIX

An Evolutionary Museum

Daniel Axelrod's Madro-Tertiary laurels would prove prickly, however. Theoretical fruition always bears the seeds of doubt. G. Ledyard Stebbins's later thinking on desert origins seems an example of this. In his 1952 paper, "Aridity as a Stimulus to Plant Evolution," he had written that genetic change and natural selection would speed up in both semiarid and arid regions. This had supported Axelrod's idea of regional desert as a fast-evolving, recent phenomenon. In a book on flowering plants published in 1974, however, Stebbins had changed his mind about desert as an evolutionary frontier.

He had maintained in his 1952 paper that slowly evolving populations "nearly all existed in environments which have remained relatively constant and continuously favorable," like "the great forest belts." But the idea that desert is an unfavorable habitat is an anthropomorphic one.

Desert is unfavorable to humans, but not to organisms that are adapted to it. By this logic, it is not the "favorable" habitat that causes slow evolution, but the "constant" one. Finding a much higher number of species per genus in the shifting patchwork of West Coast semiarid grassland and woodland than in true desert, he surmised that evolution accelerates mainly in the former, whereas desert organisms evolve more slowly, as in other "extreme habitats" such as ancient forest. In his 1974 book, Stebbins had come to regard desert not as an evolutionary frontier but as an evolutionary museum where species accumulate like specimens preserved in the dry air.

Stebbins did not present his "museum desert" scenario in opposition to Axelrod's recent desert paradigm. He illustrated it with "museum specimens" that may be fairly recent. As examples of well-adapted, stable desert bushes whose ancestors probably had evolved quickly in "intermediate" habitats, he cited "two of the principal groups of shrubby species found in arid regions of the Northern Hemisphere . . . the genus *Artemisia,* the sagebrushes, and the various shrubby genera of the Chenopodiaceae, such as *Atriplex* [shadscale], *Eurotia* [mule fat], *Grayia* [hopsage], and *Sarcobatus* [greasewood]." Noting that these shrubs have many herb relatives like mugwort and goosefoot, he postulated that they had herb ancestors in the early Tertiary:

> At that time, the prevailing climate throughout the Northern Hemisphere was mesic, with ample rainfall throughout the year, so that continuous forests existed across most of North America and Eurasia. Nevertheless, there were probably "ecological islands" of more xeric climate, resulting from the rain shadows formed by isolated mountain ranges, or from other unusual combinations of ecological conditions. As the overall climate became more arid during the Tertiary Period, these xeric islands increased in size, and new ecological islands probably appeared.
>
> The forest trees and mesic shrubs that formed the "sea" of forest surrounding these islands were already too specialized and rigid in their requirements to be able to give rise to xeric shrubs optimally adapted to the conditions on them. . . . Given this restriction, and the ease of seed transport and establishment that both now and formerly has been characteristic of families such as the Compositae and the

Chenopodiaceae, one can imagine without much difficulty that the evolution of xeric shrubs from semixeric herbs belonging to these families could take place more quickly and easily than could the evolution of xeromorphic adaptations in trees and shrubs belonging to genera such as *Acer* [maple], *Alnus* [alder], *Betula* [birch] . . . and other inhabitants of the surrounding forests.

Stebbins's "museum specimen" examples may be atypically recent for desert bushes, however. Sagebrush *(Artemisia tridentata)* is not a "desert relict" like those he had cited in his 1965 article with Jack Major. It is not an isolated species with few relatives, but a widespread one with many relatives. The possibility that sagebrush evolved rapidly as climate dried since the Miocene epoch doesn't address the question of how desert relicts like ocotillo originated.

Stebbins's 1974 book seemed tacitly to acknowledge this. It reiterated his and Major's idea that relicts originated from large "radiating complexes" that had occupied "intermediate semiarid or open regions" but had "become extinct there, having succumbed to competition from recently evolved more successful and dominant groups." Surviving only in deserts, "where the rate of speciation is rather low," the isolated relicts "persisted long after their ancestors have become extinct." They didn't evolve quickly from Miocene epoch Arcto-Tertiary woodland herbs as Stebbins thought sagebrush might have. Their abundance and diversity in North American desert imply a much longer past—a much older desert.

There were other, less tentative seeds of doubt. Axelrod's paradigm was more ascendant with life scientists than earth scientists, some of whom had long questioned whether West Coast mountains and rain shadows are as recent as generally believed. "Some scientists have tried to deduce past climates and topography from fossil evidence," wrote Jeffrey P. Schaffer, a geographer at Napa Valley College and author of many books on California geology and wilderness areas. "However, there can be problems with this method because it assumes that past climatic gradients were identical to current ones, that climatic tolerances of plants have not changed with time, that species migrate together, and that plant distributions respond to average environments more than to infrequent, extreme conditions."

Schaffer implied that the Madro-Tertiary paradigm's geological critics balanced or outweighed its biological supporters: "Daniel Axelrod in 1957, 1959, 1966, and 1980, and with William Ting in 1960, claimed to have seen changes in the composition of Nevada's Neogene paleofloras (plants of the Miocene and Pliocene epochs) that imply the creation of a rain shadow, presumably due to Sierran uplift. Mark Christensen, Herbert Meyer, and Jack Wolfe have addressed serious problems with Axelrod's data, methodology, and paleoaltitude reconstruction assumptions." And although U.S. biologists largely embraced the Madro-Tertiary paradigm, "earth-old desert" persisted south of the border like a tough if frost-sensitive plant. "Desert landscapes are the most ancient ones in our planet," wrote two Mexican scientists in 1992. "For instance, imprints of xerophilous plants are known from deposits of the upper Paleozoic (Markov, 1965). The formation of modern deserts began in the upper Cretaceous and continued at different places in the Paleocene."

The previous year, Jerzy Rzedowski had reiterated his objections to Axelrod's paradigm. He'd noted that desert plant fossils continued to be much more elusive than other kinds, and that there was "little hope that such fossils have been preserved more than sporadically."

Thus the age of the native flora of arid regions is particularly hard to estimate. Axelrod (1979) asserts that the "Sonoran Desert" of northwest Mexico and the southwestern U.S. has existed about as it is since the Pleistocene, indicating that its vegetation has evolved from enclaves of dry climate that have come into existence since the early Miocene or perhaps since the late Eocene. Nevertheless, there are no substantial reasons to assume that arid climate is such a recent phenomenon in the latitudes of Mexico, and its highly diversified xerophilic flora is more suggestive of a prolonged period of evolution, perhaps initiated in the Cretaceous itself. In support of this idea are discoveries of fossil remains of *Prosopis* [mesquite], of *Vauquelinia* [Arizona rosewood], and of *Agave* in the Eocene of the western United States, as well as of *Fouquieria* [ocotillo], *Pachycormus* [elephant tree], and *Condalia* [abrojo] in the Miocene of the same region [Axelrod 1979].

In addition, various authors (Axelrod, 1950, Raven, 1963, Wells and Hunziker, 1976) insist that *Larrea* [creosote bush] arrived in Mexico from South America in the Quaternary [ice age] or perhaps a little earlier. The southern origin of this important element of the xerophile vegetation of North America is probable, but there is no convincing evidence of its time of arrival. There exist, at the same time, clear indications that plant interchange between zones of North and South America, at least, began in older times.

Rzedowski again pointed to Mexico's wealth of endemic plants as suggestive of ancient ecosystems:

> This phenomenon is particularly spectacular in arid and semiarid zones, where endemism often pertains not only to high level taxonomic groups but also to biological forms. . . . Thus, for example, the cactus family, although originating in South America, has attained here its maximum diversity, abundance, and importance, containing around 900 species, of which more than 95 percent are restricted to Megamexico 1 [Rzedowski's term for northern Mexico and adjacent, ecologically similar parts of the United States].
>
> The family Fouquieriaceae [the ocotillo family], equally endemic to Megamexico 1 and in all probability originating here, is distinguished by the biological forms of its representatives, which are strange even among the xerophytes. No less suggestive are the variations offered by the species of *Agave,* a genus not limited to Mexico but which has its greatest taxonomic and morphologic diversity, and probably originated, here. A similar situation applies with *Yucca,* equally with *Dasylirion* [a yuccalike genus], *Nolina* [another yuccalike genus], *Krameria* [one of the relict families consisting of one genus], and various other genera.

Rzedowski's 1991 remarks did not contain much new information, although they still posed a respectable challenge to Axelrod's paradigm. But a few years earlier, oddly enough, the Alta California desert's most spectacular endemic plant had surprised U.S. scientists by failing to support some of his paradigm's assumptions about it.

TWENTY-SEVEN

The Riddle of the Palms

If sand dunes are a metaphysical essence of desert, then palm oases seem their essential complement, the moist, shady exceptions that prove the rule of fiery desiccation. A desert movie with dunes but no palms is unimaginable. One of the archetypes of this artistic convention is an oasis that geologist Clarence King described amid the Coachella Valley's sands during his 1866 desert crossing:

> Under the palms we hastily threw off our saddles and allowed the parched brutes to drink their fill. We lay down in the grass, drank, bathed our faces, and played in the water like children. . . . Our oasis spread out its disk of delicate green, sharply defined upon the enamel-like desert which stretched away for leagues, simple, unbroken, pathetic. . . . With its isolation, its strange warm fountain, its charming vegetation varied with grasses, trailing water plants, bright parterres in which were minute

flowers of turquoise blue, pale gold, mauve, and rose, and its two graceful palms, this oasis evoked a strange sentiment. I have never felt such
a sense of absolute and remote seclusion; the hot, trackless plain and
distant groups of mountain shut it away from all the world. Its humid
and fragrant air hung over us in delicious contrast with the oven-breath
through which we had ridden. Weary little birds alighted, panting, and
drank and drank again, without showing the least fear of us.

Such accounts epitomize the desert's mystique, but palm oases are not
essential to it in the way that dunes are. When there is a dry seabed or
lakebed in a desert, dunes inevitably form as the wind piles up exposed
sand. Anyone who tries camping in the ethereal-looking dune world will
experience this inexorable process unromantically as the afternoon wind
picks up and drives tiny but hard-edged quartz crystals against every
inch of skin and equipment. Oases need groundwater, which reaches
the surface only in scattered places. Much of the groundwater in today's
deserts accumulated in the last pluvial period, and is sinking.

Despite their archetypal airs, palms are not essentially desert plants,
even for oases. Most species live in tropical rain forests, and although
many do live in drier habitats like savanna and thorn scrub, they aren't
adapted to aridity as honey mesquite or catclaw acacia are. Their big
leaves transpire a lot of water, and their fibrous roots are shallow, so they
need permanently moist soil. Far from being essential aspects of desert
dunes, palm oases are rare in them. They are rare in desert overall.

Even where groundwater nears the surface in dunes, there is seldom
enough for palms. Deep-rooted shrubs like mesquite dominate the less
arid hollows of most California dunes. Sweltering on one of the state's
spectacular sand piles, the Kelso Dunes in the east Mojave, I was glad
to find oases of a sort, but not of King's romantic variety. On the lower
slopes of the 600-foot-high dunes were little thickets of desert willow,
Chilopsis, a riparian shrub related to the catalpas of suburban streets.
Tracks showed where other animals, a fox, a rabbit, had sheltered in
one thicket's thin shade—the fox quite recently judging from the fresh-
ness of its scats. But the water that let *Chilopsis* grow was many feet
underground.

The only two groves of California's native palm, *Washingtonia filifera*,

I've seen in dunes were at Thousand Palms Oasis in the Coachella Valley Preserve near Palm Springs. Even they didn't seem very archetypical, although Cecil B. DeMille filmed Bible epics there. Now protected from fire, the palms crowd the water, making the groves resemble swamps more than King's "disk of delicate green." Bushes cover even the dunes, blurring picturesque contrasts between white sands and green shade. One grove contains the only standing water I've seen in a California palm oasis, a lovely pool full of topminnows, nervous red crayfish, and blue-spangled desert pupfish, but not a "strange warm fountain" such as enchanted King. Deep in the palms' shade, it is tidily dammed, with a piped outlet. The pupfish were introduced from their native waters elsewhere in the Coachella Valley because exotic species are crowding them out.

I don't know where King's magical oasis was. Like John Van Dyke, he was prone to hyperbole so his transports may have been fanciful. His "warm fountains" and "bright parterres" emit odd echoes of William Bartram's classic eighteenth-century descriptions of Florida sinkhole springs and wet prairies. But there are Coachella Valley groves with hot springs, and Joseph Chase described oases that might have been King's minus the romance. One called Seven Palms, "partly hidden among dunes" on the windswept plain northwest of Palm Springs, had "a score or so of the trees scattered about . . . the charms of a patch of dingy salt grass, a pool of barely drinkable water, and unlimited quail, rabbits, snipe and duck."

Washingtonia filifera, commonly called California fan palm, grows mainly in steep places with rock or clay substrates that belie little surface moisture. Even at the Coachella Valley Preserve most groves occur not on the dunes but in surrounding badlands. They are spectacular, with some of the biggest palms I've seen nestled beneath sheer mudstone cliffs. Many birds visit the streambeds' trickles of surface water. Watching white-winged doves and lazuli buntings drink one fragrant morning, I saw what looked like a cloud of confetti above the rustling fronds, and then realized that a flock of white pelicans was circling far overhead. But there isn't much in the way of "bright parterres."

Groves in places like Anza-Borrego and Joshua Tree National Park tend to be even gaunter and can seem trick parodies of King's archetype. Their dark green heads look miragelike protruding from some blazing ridge, and when I reach them, I find that the trees are real but the oasis is not. Instead of a strange, warm fountain, the palms preside over scratchy mesquite thickets or alkali-crusted mud and salt grass. If surface water exists, it usually looks unappetizing, scummed with dust and bacteria, laden with dead insects.

The army officer botanist W. H. Emory probably had a more typical experience than King's when he discovered a grove in Anza-Borrego in 1846: "Here, on November 29, several scattered objects were seen projected against the cliffs and hailed by Florida campaigners, some of whom were along, as old friends. They were cabbage trees and marked the locale of a spring and a small patch of grass." Emory was glad to get the water and grass for his stock, but he didn't go on about it. Five decades later, Van Dyke was even less enthusiastic, perhaps envying King's romantic panache: "These are the so-called oases in the waste that travelers have depicted as Gardens of Paradise, and poets have used for centuries as illustrations of happiness surrounded by despair. To tell the truth, they are wretched little mud-holes."

Still, fan palms have charisma because they are the biggest, strangest plants in the Alta California desert. (Baja also has another species, *W. robusta*, but boojums trump even palms.) Palms are strange trees just to look at—the only monocotyledons, relatives of grasses and yuccas, that get so big. *Washingtonia* grows up to seventy-five feet tall with a trunk three feet in diameter. The origin of such great water-gulping plants in the continent's driest desert has always been hard to explain.

Chase imagined *Washingtonia* waving like coconut palms by vanished beaches: "Some of the groups occur along the boundaries of the sea that anciently filled the great depression which is now partly occupied by the Salton Sea, and whose beach-mark is today startlingly plain at the base of the encircling hills. Such groups, probably, represent the indigenous growths. A number more are found at higher altitudes, but of these many are known to have been planted by the present or former Indian inhabitants of the region." Other writers thought Spanish missionaries had planted groves, although this seems unlikely given their remoteness. In 1940, Edmund Jaeger declared the species "definitely known to be a native of the California desert," but characteristically did not ask how this came about.

In his 1950 article on desert origins, Axelrod cited *W. filifera* as one of the subtropical woodland species that had "closely similar fossils in the Miocene and Pliocene floras of the Mohave region and its border areas." And indeed, no tree in California seems a better candidate for the romantic role of relict from a verdant Madro-Tertiary past. Sitting in twilit palm groves, I've had a palpable sense of ghostly woodland hovering in gloom where only boulders and bushes now exist.

Of course, the same questions arise with fan palms as with other organisms that apparently once lived in nondesert but now live only in desert. James Cornett, a curator of the Palm Springs Desert Museum, observed that the present Mojave's high desert lacks wild palms because of winter frosts, but he had trouble explaining another absence: "It is less obvious why desert fan palms do not occur in the coastal climates of southern California. Winter temperatures are relatively mild and locations where permanent moisture can be obtained are numerous." Cornett guessed that "relatively cool summer temperatures and/or frequent fires" excluded palms from the coast by giving other tree species a competitive advantage around water. But this explanation seems tenuous. Inner coastal valleys get almost as hot as desert in summer, and fan palms are resistant to fire. In the desert they can benefit from fire, since it clears the soil for their seedlings.

In the late 1980s, Leroy R. J. McClenaghan and Arthur C. Beauchamp, two San Diego State biologists who wanted to study genetic variation in wild plants, chose fan palm groves in Anza-Borrego as a subject. Regarding *Washingtonia filifera* as a relict of Madro-Tertiary woodlands, they made "three a-priori predictions about the genetic structure of this species." They expected the isolated palms to have less genetic variability than more common and widely distributed plants, since variability dwindles in relict populations as they become smaller and more scattered. They expected larger groves to have more variability than smaller ones, since less would have been lost. And they expected that each isolated grove would show a lot of variability in relation to other groves, since "genetic drift" would have occurred during their isolation.

Only the first prediction proved right. The palms did have less genetic variability than commoner plants. But big groves did not have more genetic variability than small ones, and the level of variability from grove to grove was not as high as anticipated. This was surprising for a species that supposedly had been widespread in the Madro-Tertiary Geoflora at least since the Miocene, and then had retreated to scattered oases as climate dried and cooled. It suggested that fan palms had come to occupy today's desert in a different way:

> A more plausible scenario would be one in which existing palm popula- tions in the Colorado Desert are the products of seed dispersal from a source population having reduced genetic variability. It has been sug- gested that climatic changes may have completely eliminated fan palms from the Colorado Desert and restricted the species to small refugia populations in Baja California (J. Cornett, pers. comm.). The subsequent movement of seeds into southern California from a refugium popula- tion in northern Baja California would have resulted in these colonizing populations also having low variability and high genetic similarity because of their common ancestry. Gene flow among these colonizing populations would have reinforced their genetic similarity.

Birds, coyotes, humans, and other consumers of the small but sweet (tasting like dates) palm fruits presumably would have carried out the gene flow.

Cornett originally had agreed with Axelrod about fan palm origins but had changed his mind. In 1989, he cited "four lines of evidence" against *Washingtonia filifera*'s supposed Madro-Tertiary relict status. First, he maintained that the species has never been as widely distributed as it is now:

> There does not, in fact, appear to be any fossil evidence indicating that the species was once widespread in the Mojave Desert. Axelrod (pers. comm.) with one exception has retracted his earlier assertions that the fossils in question could be assigned to the genus *Washingtonia*. The one exception is a fossil collected near Wikieup, Arizona, and deposited in the collections of the Museum of Paleontology at Berkeley. Axelrod

believes it can be classified as *Washingtonia*. However, the specimen could not be located at the Museum, and thus could not be examined by the author. An examination of additional Axelrod palm fossils by the author failed to reveal any specimens that displayed thorns on the petioles [leaf stems], an important characteristic of the genus, and therefore none could be classified as *Washingtonia*.

Second, instead of declining in numbers as might be expected of a relict species, *Washingtonia* was increasing, from approximately 17,700 individuals in 1961 to 23,266 in 1987. Third, instead of contracting in range like a declining relict species, *Washingtonia* was expanding, establishing new wild populations in Death Valley as well as in Nevada and Arizona. Fourth, as the two San Diego State biologists had determined, instead of having large genetic diversity among scattered populations, *Washingtonia* had little diversity, suggesting that the populations had diverged fairly recently. "These four lines of evidence," wrote Cornett, "all point to a recently-evolved, invasive species, not a relict. It seems most likely that the genus *Washingtonia* first evolved in Baja California sometime after the peninsula broke away from mainland Mexico approximately 4.5 million years ago. Today, the two species in the genus, *W. filifera* and *W. robusta*, occur together only in Baja California, suggesting this is the geographic origin. Had the genus been present before the peninsula broke away, one would expect it to be represented on the mainland, which it is not."

Cornett thought that *Washingtonia* might have evolved from *Brahea*, a related palm genus that occurs today in both mainland Mexico and Baja California. He speculated that "the glacial episodes of the Pleistocene" might have influenced the evolution of a "cold-tolerant palm" like *Washingtonia*, and that it might have "appeared within the present boundaries of the United States no earlier than the end of the Illinoian glacial episode." In 1991, Cornett further speculated that, like opossums and armadillos, "the desert fan palm is another example of a species with tropical affinities that has extended its range northward during this period of global warming."

Cornett did not speculate as to the significance for the desert's pre-Pleistocene past if *Washingtonia* was not a Madro-Tertiary flora relict. Of course, the possibility that one species is not such a relict did not necessarily challenge Axelrod's desert origin ideas. Whether or not Axelrod's Madro-Tertiary fossil palms are *Washingtonia* or *Sabal*—the other genus Axelrod identified in desert deposits—they did exist along with the other fossil trees he found in today's desert. A nonrelict *Washingtonia* might just be an exception to the Madro-Tertiary rule. Axelrod did not respond in print to Cornett's ideas, suggesting that he found them unchallenging.

William Spencer, a naturalist, did respond to Cornett's ideas, challenging most of them. Spencer argued that Axelrod's fossils could indeed be *Washingtonia*, since fossil palms are hard to identify and many *Washingtonia* petioles lack spines. He argued that *Washingtonia* might not have increased since the 1950s, since early palm counts were haphazard. He argued that the low genetic variability of Anza-Borrego *Washingtonia* might not prove nonrelict status, since broadcasting of seeds by humans and wildlife could have homogenized the population. And he maintained that *Brahea,* the Baja fan palm genus, is too different from *Washingtonia* to be an ancestor.

Spencer didn't emphasize Axelrod's paradigm in his defense of *Washingtonia*'s ancient relict status. His opposition to Cornett's ideas focused more on the assertion that *Washingtonia* had recently spread from Baja to Alta California than on Cornett's implicit denial that *Washingtonia* grew in Miocene Madro-Tertiary woodlands. Spencer seemed more interested in affirming that *Washingtonia* had been in southeastern California "since the beginning" than in determining when that beginning might have been.

Ironically, one possible interpretation of Cornett's ideas also implies that fan palms could have been part of the California desert flora virtually since the beginning. If *Washingtonia* did not evolve from a Madro-Tertiary woodland species, but arrived from the south, it could be a product, even if a fairly recent one, of Jerzy Rzedowski's ancient Mexican desert.

TWENTY-EIGHT

Bushes and Camels

The same year that the two San Diego State biologists published their analysis of *Washingtonia* genetics, another biologist published yet another possible explanation for desert fan palm's limited distribution—that its main dispersers are extinct: "It is likely that contemporary rare desert trees with very localized distributions and fleshy fruits (e.g. the desert palm, *Washingtonia filifera,* which occurs in tiny groves in the Sonoran desert and has canid dispersed seeds at present) could become very common if once again serviced by a wide-ranging megafaunal dispersal agent such as a camelid." The biologist was Daniel Janzen, who became prominent in the 1980s for his studies of the relationships between vegetation and megafauna, the diverse assemblages of large animals that inhabited the Americas until ten thousand years ago. Although best known for his work on tropical dry forest in Central America, Janzen also studied deserts:

There are many books and general treatises on the vegetation and vegetation types of the deserts of north central Mexico and the southwestern United States. None give consideration to the role played by large
herbivorous vertebrates in shaping individual plants or their arrays. . . .
Papers on cactus ecology classically ignore the Pleistocene (and earlier)
megafauna, as do studies of spacing of desert plants. . . . In like manner,
detailed discussions of the evolution of the biologies of the Pleistocene
and pre-Pleistocene megafauna rarely consider that virtually the entire
flora of large desert plants must have been continually under selection
for defenses against these mammals. The camel has the largest gape of
any extant ruminant and eats very thorny vegetation. The relationship
of such a pair of traits is probably not evolutionarily fortuitous, and if it
is, it still needs consideration to understand ecological fitting of camels
to deserts.

Finding Miocene and Pliocene camel, mastodon, and horse fossils at
Red Rock Canyon and Barstow in 1915, John Merriam had surmised that
southeast California must have been grassland or savanna to support
so many large mammals. In a lengthy 1967 paper on the extinction of
American megafauna, Daniel Axelrod blamed its disappearance partly
on the replacement of woodlands and grasslands by regional deserts. But
Janzen's research on African deserts led him to believe that American
ones could have supported the equivalent of giraffes, elephants, and
diverse other herbivores as well as big carnivores such as the African
lion, which also inhabited California: "There were four genera of cam-
elids in the western half of North America at the close of the Pleistocene,
and they probably treated the North American deserts just as the contemporary African camel does its deserts. Camels move long distances
among local wet sites, eat fruits, and defecate seeds, and range from
Kenyan thorn forests to the driest deserts."

Likewise, Janzen's research on large herbivore behavior in relation
to desert plant features such as thorns and poisonous resins suggested
that those features played a role in defending the plants, and thus had
evolved in response to browsing more than other factors:

In the absence of contemporary wild medium-sized to large herbivores,
it has been fashionable to try to understand the spininess of arid-land

plants largely in the context of their interactions with the physical environment. Spines on desert plants "were probably developed in the first place as a response to the dry atmosphere. . . . Furthermore, thorniness is most highly developed in the most arid deserts, exactly where large grazing animals are rarest." Such a comment needs to be paired with the alternative view that the scarcer the perennial vegetation, the better protected it must be to survive. . . . It is not hard to imagine how browsing megafauna could select for arborescent lilies or botanical hedgehogs.

Janzen cited a Mexican tree as a contemporary example: "*Acacia farnesiana* clearly has a memory of browsers that were—e.g. leave it alone and the thorns are short and the leaves long past them, browse it with a pair of clippers and the next branches to be produced have wicked long thorns and shorter leaves among them—something I saw happen also with some native acacia in Morocco being browsed by camels and goats—below the browse line, fierce thorns, above the browse line much reduced thorns."

Janzen was particularly fascinated by nopal, the tall, branching prickly pear that, two centuries before, had given the Jesuit Miguel del Barco inklings of biotic change. Noting that, except for humans, coyotes, and various insects, few animals now feed on the sweet, juicy nopal fruits, Janzen wondered how they could have evolved if not in response to browsing by large mammals such as camels, mastodons, and ground sloths. What other creatures would have been strong and thick-skinned enough to plow into "nopaleras"—thorny thickets of cactus and other plants such as mesquite, acacia, and creosote bush—to reach the fruits? What others would have had mouths and stomachs tough enough to chew and digest the prickly pears? What other creatures would have broadcast the seeds widely enough in their dung to make *Opuntia* cactus so successful?

"Then why the bright colors of *Opuntia* fruits? The traditional view of mammal color vision is that it is restricted to primates, tree shrews, and ground squirrels. However, a series of color choice tests with wapiti (*Cervus canadensis*) show clearly that they can distinguish orange from a variety of other colors, and cones have been located in the retinas of white-tailed deer and wapiti. I view the brightly colored large fruits of *Opuntia* as circumstantial evidence that at least some of the large herbivorous megafauna used color vision in food location."

Janzen thought desert plant assemblages such as nopaleras would have provided a megafaunal smorgasbord, offering not only prickly pears but a variety of other conspicuous fruits—yellow mesquite and acacia pods, banana-like yucca and agave fruits—plus seeds, leaves, stems, roots, and even the cactus pads themselves. He cited the presence in a Shasta ground sloth's dung from an Arizona cave of roots, stems, seeds, flowers and fruits of ephedra, globe mallow (a shrubby okra relative), saltbush, mesquite, agave, yucca, and nopal.

Janzen also posited a megafaunal origin for the vegetatively reproducing jumping cholla cactus: "Jumping cholla may well be the nastiest of the world's burrs. When cacti break up at the stem joints through rough (or not so rough) treatment by contemporary herbivores eating or trampling stems or fruits, they are probably displaying a response selected for earlier by much more brutal treatment. It is easy to imagine the early evolutionary stages of jumping cholla as simply the spines on cactus pads lodged in thick skin on large feet."

In another article, Janzen theorized that megafauna also might have affected smaller animals: "During recent fieldwork in Kenya (dry season, February 2–9, 1974), I saw not a single lizard or snake, and only one turtle, in about 1,000 miles of rural roads and 4 days of close scrutiny of four national parks ranging from 3,000 to 10,000 ft elevation. . . . Covering similar terrain and vegetation during the dry season in Mexico, Costa Rica, Colombia, or Venezuela, I would have seen hundreds of lizards and some snakes, with the same type of searching." Janzen then conducted an informal survey of African versus American big game and reptile abundance, estimating that Africa's reptile biomass was roughly 10 to 15 percent of its large mammal herbivore biomass, while America's large mammal herbivore biomass in warm deserts and southward was 10 to 15 percent of its reptile biomass.

Janzen thought megafauna might depress reptile biomass in two ways. First, availability of carrion even during times of reptile scarcity could maintain large populations of raptors, small canids, and other habitual predators on reptiles. Second, big game herds could degrade reptile breeding and feeding habitat by browsing, trampling, and otherwise disturbing vegetation.

He thought herbivorous reptiles would be particularly vulnerable, not simply because of their dependence on plant food, but because of their metabolic requirements: "The ease with which leaf eating should evolve in a lizard fauna should be decreased as predator intensity increases, since it appears that very long periods of basking are an integral part of the digestive behavior of leaf-eating lizards. . . . The total absence of a foliage-eating *Iguana* or *Ctenosaura* [ground iguana] analogue from the African tropics is very conspicuous." Janzen noted that African leaf-eating lizards are scarce not only in forests and savannas but in deserts: a study in the Kalahari recorded 1.2 percent of lizard gut contents to be plant matter, compared to 8.3 percent in North America. And the Kalahari has more lizard species than North American desert, although many are small and nocturnal.

Are California's abundant desert reptiles beneficiaries of the American megafaunal extinction? This might seem more likely if the megafauna had died out here ten million years ago instead of ten thousand. We know that diverse reptiles very like today's desert ones coexisted with camels and mastodons in the Anza-Borrego region several million years ago, even though most of the region's known plant fossils aren't desert ones. Yet the fact of our desert reptiles' success remains. Might there have been "earth-old" deserts wherein extreme dryness or other factors released distant ancestors of chuckwallas and desert iguanas from megafaunal pressures long enough for them to evolve their unusual vegetarian habits?

Janzen did not ask such questions, confining himself to the Pleistocene epoch and recent times. He thought that many desert plants might have reached North America after the Central American land bridge formed three million years ago: "Cacti are widely believed to be of South American origin (A. Gentry, personal communication). If so, their original evolutionary interactions would have been first with the independently evolving South American megafauna, rich in large animals such as ground sloths, glyptodonts, and toxodonts, and later with the North American Pleistocene and pre-Pleistocene megafauna as the cacti moved northward (probably as seeds in the guts of megafauna)."

Many North American desert plants such as ocotillo and nolina are endemic, however, and even cactus origins are uncertain. And while desert plant fossils remain scanty in North America, fossils of big mammals that are known to adapt well to deserts, such as equids and camels, are abundant. Some Miocene American camels had giraffelike, elongated front legs and necks that could have been used for browsing in the crowns of *Washingtonia* palms, Joshua trees, or tall cactuses.

Of course, a fossil camel's long neck does not prove that it fed on such plants. Most desert plants are short. Even if they did browse on tall ocotillos and Joshua trees, pre-Pleistocene camels might have interacted with such plants not in large regional deserts like today's but in the thorn scrub, savanna, and woodland mosaics of Axelrod's Madro-Tertiary Geoflora. "The plant-megafaunal interactions within the nopalera cannot be viewed in isolation from other habitats," Janzen wrote. "For example, it is well appreciated that during the Pleistocene glaciations a much more forest-like vegetation covered what is presently desert and semidesert in northern Mexico and the southwestern United States. . . . It is easy to visualize a herd of gomphotheres [mastodons] ranging into nopaleras to eat *Opuntia* fruits in the summer, moving into the oak forest to eat acorns in the fall, and then back into the nopalera to eat *Opuntia* pads in the winter." Still, the idea that large mammals shaped desert plant biology implies lengthy coevolution during times of high stress from aridity. Large herbivores would have treated woodland and savanna plants as brutally as they did desert ones, but those well-watered plants did not evolve such extreme defenses.

When I asked Janzen about this, however, he replied that growing in full sunlight can influence plants as much as growing in dry climate: "Ocotillo and cacti, for example, don't need the dryness near as much as they need the sun that comes with the dryness and ensuing non-other canopies shading them. This means that desert-old plants can grow very well in non-desert places (e.g., rocky ridges, beach cliffs) and have desert-old life forms and co-evolve with large mammals etc." That seemed compatible with Axelrod's idea that desert plants evolved in scattered dry nondesert sites. But Janzen was too busy with Costa Rican conservation emergencies in 2008 to further pursue his ideas of two decades earlier. Prehistoric megafaunal effects on desert evolution remain largely unstudied.

Once I got the idea of megafaunal interaction into my head, though, it was hard to see the desert as I had before. When I took a walk on Cima Dome in the Mojave National Preserve one windy afternoon, the extraordinarily dense and tall Joshua tree woodland there seemed to shout: "Camels! Giraffe-camels!" Each branch of spiky yucca leaves clutched its crown of soft white blossoms as though holding them as far as possible from some ghostly browser. It was easy to imagine fifteen-foot Miocene camels sauntering through and nipping a cluster here and there.

As Janzen suggested, such a scene might not have been in a regional desert like the present. Cima Dome is one of the "edaphically arid" environments that Axelrod elected for predesert evolution of desert plants—a granite dome with thin, fast-draining soils and many bare slabs and boulders. At over five thousand feet elevation, with bunchgrass and juniper as well as Joshua trees and cactuses, it is more like Pleistocene woodland of ten thousand years ago than today's scanty creosote bush and burroweed scrub. Even so, the idea of giraffe-camels feeding on the Joshua trees seemed to imply an older, more obscure desert past than Axelrod's. And the possibility of such a past loomed larger in the years after Janzen's papers.

TWENTY-NINE

Axelrod Askew

It is hard to imagine even camels thriving in some of California's present desert. Fortynine Palms Canyon on the north edge of Joshua Tree National Park is an example, a landscape of granite boulders like Cima Dome, but without Joshua trees, junipers, grass, or much else in the way of giant mammal fodder. Even spring wildflowers are sparse among the creosote bush and barrel cactus. The palm oasis at the top provides water for bighorns and deer, toads and tree frogs, but it is the usual *Washingtonia* grove, a crack in the rocks with a dusty trickle at the bottom. Still, the canyon does feature one relatively gigantic inhabitant. Despite its sparse vegetation, it has a highly visible chuckwalla population.

Perhaps this is because the granite there is particularly large grained and easy for scrambling about on, or just because the big lizards are used to hikers. Anyway, the canyon was the scene of the most dramatic lacertilian dalliance I've come across. While walking up the trail late one spring afternoon, I glimpsed something bluish black protruding above an upended boulder. When it began to bob up and down, I saw that it was the swollen, pop-eyed head of a male chuckwalla. Then a smaller greenish gray, wrinkled head appeared as if in response to the first's push-ups—a female. They posed there in profile awhile, like pre-Columbian demigods on a stela.

Suddenly, the male moved up farther and, turning his body broadside to the female, displayed red and yellow spangles that looked iridescent as fire opal in the late sunlight. I found this spectacular for a usually drab species, and so did the female. When I climbed to a better vantage point, I could see that she was quite excited. She kept nosing at his tail, and twice got so worked up that she crawled on his back and rode him around briefly. She paid more attention to him than he did to her. After making his spectacular color display, he just kept doing push-ups, propped on a protruding bit of rock like a heraldic beast on a shield.

He eventually condescended to nose at *her* tail, but playing hard to get can backfire. She rebuffed him, scurrying away down the boulder's side, and it appeared beneath his dignity to chase her. He dawdled there awhile as though trying not to seem disappointed. Then while the sun was setting he hurried in the opposite direction, probably toward the safety of his rock lair.

I could see why his tail had excited her. It was impressively long, plump and sinuous, colored a pristine creamy yellow that contrasted strikingly with the blue-black rest of him. It called to mind another Lewis Carroll nonsense verse:

How doth the little crocodile
Improve his shining tail,
And pour the waters of the Nile
On every golden scale!

That reminded me of something I'd noticed about the California desert fossil record. Crocodiles like those of today's tropics have existed since the dinosaur age. California probably was too cold for them in the past few million years, but before that the climate was often warmer. At least three tropical dugong species left fossils in mid-Miocene marine deposits. If Madro-Tertiary California was as low lying and well watered as Daniel Axelrod maintained, it should have had many wide, slow rivers where fish were abundant and large herds came to drink—good habitat for crocodiles and for preserving crocodile bones. Their fossils abound elsewhere in North America. But I've seen no mention of them in California desert fossil deposits, which suggests that it was not such a low-lying, well-watered region.

Apparent crocodile absence is a minor anomaly in the Madro-Tertiary paradigm, one that Axelrod could have swept aside easily. After all, a few alligator-like fossils from just after the dinosaur age are known from the El Paso Mountains; perhaps later crocodiles simply haven't turned up yet. Still, it is far from the only anomaly, and his reticence on desert evolution after the mid-1980s could have had another cause besides his paradigm's ascendancy with biologists.

UC Davis botanist Michael Barbour made a good point in favor of his friend when he wrote, "During an era when most scientists became more and more specialized, Axelrod retained an ecosystem-level focus and curiosity. He asked, and answered, large questions." But Axelrod's virtuoso self-reliance had its downside. Rumblings of geological discontent had continued to grow, and a flurry of papers published in the 1990s may have left the volatile octogenarian speechless. Based on new techniques and concepts, they proposed ideas wildly at variance with his desert origin paradigm's basic premise that California's present high mountains began rising in the Pliocene epoch 5 million years ago.

Scientific paradigms undergo life cycles—from slow beginnings as skepticism and rivalry play out, to sudden ascendancy as the balance tilts in their favor, to surprise upsets as new scientific generations arise. The scientific generation that arose in the 1990s was impressively new. Their papers always had multiple authors and they covered subjects so technical and specialized that the authors sometimes seemed nonplussed by their own conclusions.

A 1996 paper on West Coast mountain origins in *Science,* one of the two leading journals, had nineteen authors. In four densely documented pages, it briskly knocked Axelrod's basic premise askew by suggesting that California's major mountains arose long before the Pliocene and have been high ever since: "The Sierra may have maintained or even lost elevation in the late Cenozoic as paleofloral and geomorphic arguments for uplift of the western United States have recently been questioned." Using tectonic theory and new chemical techniques for analyzing fossil leaves and igneous rocks, the authors proposed startling new estimates of ancient elevations. They estimated that the Sierra Nevada rivaled today's Andes in the late dinosaur age, rising to 18,000 feet above sea level overall, and that, although it lost elevation during the past 65 million years because of erosion and continental crust thinning, it was still an average 9,000 feet high 5 million years ago.

A year earlier, an article in another prestigious journal, *Geology*, had suggested that the Mojave region and the rest of southeastern California had "collapsed" some 24 million years ago when the Pacific and North American tectonic plates moved apart, thus stretching and thinning the continental crust. This raised the possibility that today's western landscape of arid interior basins in the rain shadow of coastal mountains might have been the landscape of 20 million years ago as well.

In 1997, another *Science* article whose four authors included Jack Wolfe—cited by geographer and author Jeffrey Schaffer as among Axelrod's main critics—challenged his basic premises about fossil floras:

> Terrestrial plants are generally regarded as highly responsive to environmental changes, and thus fossil plants offer one of the best methods for inferring paleoenvironmental parameters. In one approach, the environmental tolerances of a fossil species are assumed to be the same as those of its nearest living relative. Because in this method it is assumed that plants do not evolve by adapting to different environments, conclusions based on this method have been interpreted to indicate that most of the Basin and Range Province of Nevada was at low altitudes (< 1 km) [under 3,000 feet] until < 5 million years ago, when uplift was inferred to have started which resulted in the 1 to 1.5 km present-day mean altitudes of the basins (based on Axelrod's publications in University of California Publications in Geology from 1958 to 1992).

The paper posited startlingly different fossil plant "parameters" than Axelrod's. It estimated, from "multivariate analysis of leaf physiognomy" of twelve mid-Miocene floras, that the Great Basin region stood a stagger-ing 12,000 feet above sea level in the Miocene epoch 16 million years ago, almost three times its present altitude: "Paleobotanical evidence sup-ports the hypothesis that Mesozoic thrust faulting and crustal thicken-ing built a high terrain in what is now the Basin and Range province. . . . Geophysical observations combined with theoretical considerations of the region to the south of our study area suggested high altitudes at 20 Ma [million years ago] and a subsequent collapse; an increasing body of data and interpretations argue for high altitudes during the Tertiary in much of western North America."

Axelrod's fossil classifications also came into question. According to Schaffer, "Howard Schorn, a very meticulous paleobotanist, reevaluated Axelrod's fossil plant identifications, and found that most of the plants were misidentified. When they were properly identified, they are found to be plants of former highlands, not lowlands, for Nevada."

Schorn also questioned fossil identifications in other states. Axelrod had dated a deposit in eastern Oregon, the Alvord Creek flora, in the Pliocene epoch, and he had identified leaf fossils there as belonging to species of rose, madrone, ceanothus, and serviceberry (*Amelanchier*). Since species of these genera occur on the California coast today, this implied a low-altitude landscape. In 1994, Schorn and Nancy L. Gooch dated the Alvord Creek flora to the late Oligocene or early Miocene, some 15 million years earlier, and attributed Axelrod's rose, madrone, and ceanothus leaves all to serviceberry. Most western serviceberry spe-cies today grow at higher altitudes, suggesting that Axelrod had gotten the elevation as well as the date wrong.

In 1998, the year Axelrod died of heart failure, a paper in the other leading journal, *Nature*, estimated ancient elevations similarly to the *Science* papers:

> We conclude that the Sierra developed an Andean-scale topography
> between 185 Myr (the youngest marine strata preserved in batholith
> wallrocks) and 70–80 Myr (the minimum age of deep incision) coinci-
> dental with crustal thickening via arc magmatism [massive intrusions
> of molten rock caused when crustal plate motion shoves volcanic island
> arcs against continents] in the Sierra and contractional deformation to

the east. By 70–80 Myr ago, both the San Joaquin and Kings drainages had deeply incised the western flank of the range, after which we infer gradual reduction of mean crestal elevation from 4.5 km to 2.8 km.

Again, far from rising abruptly to its present overall height of 9,000 feet in the past 3 million years, the Sierra Nevada shrank slowly from a mean height of 15,000 feet in the past 65 million years.

In response to a laudatory obituary by Michael Barbour, Schaffer somewhat severely summed up the case against Axelrod in 1999. Referring to his work in identifying fossil pollen, Schaffer wrote: "Because he believed that the Sierra Nevada and Great Basin were low landscapes until late Cenozoic time, he identified all his specimens as low-elevation analogs. Professor Roger Byrnes (among others) once told me that about ninety percent of Axelrod's pollen identifications were wrong. Unfortunately, most botanists don't realize this and they repeat his errors. Ax's geology was atrocious. . . . It will take decades or longer to purge his voluminous imaginary species, imaginary landscapes, and imaginary climates from the field of paleoecology."

The new scientific generation continued to hammer Axelrod's paradigm in the new century. A 2002 article estimated ancient elevations by analyzing silicate isotopes in Miocene and Pliocene volcanic ash:

> Volcanic ashes currently exposed in the rain shadow of the modern Sierra Nevada of California show no indication of large scale late Cenozoic surface uplift of the Sierra and corresponding regional rain shadow development. Rather smectite isotope data tentatively suggest that elevations may have decreased over this time by as much as 2000 m toward the southern end of the range and 700 m in regions farther north. This suggests that the modern rain shadow cast over the western Basin and Range has been in existence since pre-Middle Miocene and that the Sierra Nevada has been a prominent orographic barrier since before this time.

If the southern Sierra Nevada stood some 6,000 feet higher 20 million years ago than it does now, that would have created an impressive rain shadow indeed. And what would have grown in that shadow? Presumably not the oak and pine woodlands whose fossils Axelrod dug

up in the Mojave. According to his critics, those woodlands would have
been growing in mountains at higher elevations. Presumably an arid-
adapted vegetation would have grown in the rain shadow, perhaps one
not so different from today's Mojave Desert.

Yet another multiauthored paper based on new techniques suggested
something of the kind in 2004, although it was about Oregon's desert
instead of California's. The paper compared 30-million-year-old Oligocene epoch "paleosols," fossil soils, with ones from the last ice age, and
found them "surprisingly similar." In successive strata of ice age paleosols, trace fossils of earthworms alternated with trace fossils of cicadas,
implying shifts between grassland and sagebrush scrub. (Earthworms
are rare in dry scrubland soils: cicadas feed on woody plants.) In the
30-million-year-old paleosols, earthworm traces also alternated with
cicada traces. The paper also mentioned possible traces of desert succulents (Cactaceae, Euphorbiacea) and shrubs (such as saltbush, *Atriplex*)
in the Oligocene soils.

A 2006 *Science* article estimated the northern Sierra's early Age of
Mammals elevation by analyzing hydrogen isotopes in Eocene river
clays exposed by gold mining on the Yuba River. The estimated elevation was not as high as the southern Sierra's, but it was still pretty high,
around 6,500 feet: "The data, compared with modern isotopic composition of precipitation, show that about 40 to 50 million years ago the
Sierra Nevada stood tall (plus or minus 2200 meters) a result in conflict
with proposed young surface uplift by tectonic and climatic forcing but
consistent with the Sierra Nevada representing the edge of a pre-Eocene
continental plateau."

The new "old desert" paradigm is not confined to North America.
Another *Science* article published in 2006 reported dune deposits in the
Sahara Desert from the late Miocene epoch, 7 million years ago. An
article in the same issue estimated the rise of the Tibetan Plateau, which
isolated central Asia from rainfall, at about 35 million years ago, tens
of millions of years earlier than was previously thought. Other studies
have pushed back the ages of South American deserts like the Atacama
almost as far. Global desert may have been less of a climatic accident and
more of an old earth feature than Axelrod thought.

THIRTY

Paradigms Postponed

It seems perversely counterintuitive that an Andean-size massif domi-
nating the far west at the Age of Reptiles' end should have persisted
to become the present Sierra and Great Basin ranges but have left only
esoteric traces of its existence throughout the Age of Mammals. The West
Coast's jagged peaks and steep canyons *look* so young compared to the
hulking Appalachians, which are supposed to be truly old mountains.
Yet *Science* and *Nature* published no letters in opposition to the articles
proposing this reversal of long-accepted notions. Subsequent issues carried arguments about climate change, animal rights, drug abuse, human
evolution, science funding—not about Miocene rain shadows.

One 2004 *Geology* article did offer some evidence of newly rising far-
west mountains: "Recent geologic data have polarized the debate about
whether the Sierra Nevada underwent late Cenozoic (ca 10 Ma to the

present) uplift. The debate has suffered from a lack of landscape erosion rates particularly from the rugged southern Sierra Nevada. We report new erosion rates that link many of the previous data sets and inspire new conceptual models of the late Cenozoic topographic evolution of the range." The three authors determined the "new erosion rates" by dating the sediments that have accumulated in caves of southern Sierra Nevada marble deposits, and thus estimating how long the Sierra's rivers had been eroding the landscape. They thought that the southern Sierra had begun to rise between ten and three million years ago "in a pattern that steepened the gradients of westward flowing rivers. These rivers responded in a wave of incision that propagated upriver from the edge of the Central Valley, deepening preexisting canyons. Incision in the marble belt reached its maximum rate as the wave of rapid incision passed from 5 to 2 Ma."

A Sierra Nevada that rose rapidly between five and two million years ago might reaffirm the conventional view that the West Coast was low lying before then. The *Geology* authors didn't think so, however. They agreed with the *Science* and *Nature* articles:

> Low temperature geochronology studies suggest that the southern Sierra Nevada had high elevations and relief as early as the late Cretaceous, when the range was an active volcanic arc. The . . . minerals east of the Sierra Nevada crest suggest a persistent rain shadow throughout the Miocene (Poage and Chamberlain, 2002) indicating that elevation was higher then as well. These data have been used to argue for a monotonic decline in mean elevation and local relief through the Cenozoic, implying no recent uplift. . . .
>
> We note that these data are not necessarily at odds. Tectonically driven rock uplift . . . would rejuvenate incision in pre-existing canyons, resulting in further flexural isostatic uplifts.

Just because the western mountains started getting higher quickly some five million years ago, in other words, it doesn't mean that they haven't been high since the dinosaur age.

But if Daniel Axelrod was askew, he was not overturned. The lack of published disagreement with the *Science* and *Nature* articles didn't necessarily signal agreement. As Jeffrey Schaffer complained in 1999, biologists continued to espouse the Madro-Tertiary paradigm. In an excellent 2008 book on California desert ecology, botanist Bruce Pavlik summarized the paradigm better than Axelrod:

> Prior to the rise of the Cascade-Sierra-Peninsular chain, western North America was a low undulating plain dominated by forests of conifers and broad-leaved hardwoods. Twenty-five million years ago, the coast ranges did not exist, and the ancient Pacific Ocean lapped a continental edge along the east side of what is now the Central Valley. The Nevadan hills ran north to south along this coast, eroded remnants of a once high range of sedimentary mountains. With an estimated elevation of only 3,500 feet, the hills were too low to block storm fronts, and rainfall was abundant, reliable, and evenly distributed across the continental interior. To the south, the influence of subtropical climates created oak and palm forests, rich pine and juniper woodlands, and thorny shrublands with exposure to seasonal drought and high temperatures, but there were no desert ecosystems at this time. The ancestors of modern desert species were, however, already establishing lineages under these semiarid conditions.

Axelrod's paradigm is not, after all, imaginary, but a genuine theory based on a great deal of informed observation and evidence, including a fossil collection that is a major scientific contribution in itself. The circumstance that his intellectual ingenuity and powerful, sometimes abrasive personality contributed to its ascendancy was not an unusual one in the real practice of science. If he was an occasional obstacle to further inquiry, he was an incitement to it as well. Whether or not they all prove accurate, Axelrod's ideas have value as the product of resourceful, perceptive exploration, just as do Miguel del Barco's, Ivan Johnston's, and Jerzy Rzedowski's.

In any case, the new western geology did not have much to say about coastal mountains south of the Sierra Nevada. Even if the Sierra and Great Basin were much higher much earlier than once thought, that doesn't necessarily mean the Transverse and Peninsular ranges now rain-shadowing California's southern Mojave and Sonoran were. Lowland Madro-Tertiary vegetation might have covered what is now southeast California even if high mountains loomed to the north.

Early paleontologists' vision of Age of Mammals southern California as a megafauna-rich plain covered with grass, savanna, or woodland remained compelling, and there was even some new fossil evidence for it. In 2006, the paleontologists who described the Pliocene-epoch Anza-Borrego region as "a lowland of meandering rivers and riparian streams" announced a discovery that questioned an earlier challenge to Axelrod's paradigm, although it challenged some of Axelrod's own fossil identifications in the process. They reported the finding of "permineralized monocot wood, morphologically and anatomically identical to modern fan palm, *Washingtonia filifera*" in Pliocene deposits at least three million years old there:

> This may represent the first documented fossil *Washingtonia* sp. within California. Previously (Axelrod, 1939, 1950 b) specimens of this taxon have been assigned erroneously to the genus *Sabal* based on palm frond impressions. This leads to paleobotanic confusion, since it is difficult to identify fossil specimens accurately from their leaf stem impressions alone. The leaf and associated fossil wood material from Anza Borrego, however, is indistinguishable from the genus *Washingtonia* sp. (based on anatomical pore structure morphology) and does not resemble *Sabal* palm. It is unlikely that *Sabal* (usually referred to the species *Sabal miocenica*) and *Washingtonia* sp. co-existed in Anza Borrego during the Pliocene. *Washingtonia* sp. favored temperate conditions and had a much wider distribution.

The Anza-Borrego paleontologists didn't mention James Cornett's 1989 theory that *Washingtonia* is a desert genus that only spread from Baja to Alta California after the ice age, but they seemed to negate it. They said in effect that *Washingtonia* was originally a palm of low-lying Madro-Tertiary riparian woodland and that the late Miocene and Pliocene palms that Axelrod and others had identified as *Sabal* were actually *Washingtonia*.

Other fossils besides plants and land mammals might suggest a low-lying southern California coast before three million years ago. A chain of marine deposits full of whale, shark, and pinniped bones that runs from the southern Sierra foothills down through Baja shows that much of today's coastward land was undersea during the late Oligocene and Miocene epochs. There might have been a wide coastal plain inland.

On the other hand, neither riparian tree fossils nor marine fossils rule out the possibility of high coastal mountains. Riparian forest grows along high-altitude streams if the climate is warm enough. I once rode a train across Mexico's Sierra Madre just north of the Tropic of Cancer from Los Mochis to Chihuahua. After climbing the steep subtropical canyons that open onto the narrow coastal plain, we emerged into an oak, madrone, and pine forest with cottonwood-bordered streams like coastal California's today, but thousands of feet higher. If California's mountains stood above nine thousand feet in the mid-Miocene, when climate was tropical, such a forest might have grown high on them as well.

If *Washingtonia*'s ancestors were as widely distributed in California Miocene and Pliocene woodland as the Anza-Borrego paleontologists suggest, they also might have grown at higher elevations. Most fan palm groves occur in steep, rocky terrain today. The only apparently natural old grove outside California, in the Kofa National Wildlife Refuge in western Arizona, grows not in a canyon bottom as might be expected but along spring channels near the top of overhanging cliffs. I searched most of a morning before I finally looked up and saw the wind-tossed main grove, which somehow reminded me of Marcel Duchamp's vertiginous painting, *Nude Descending a Staircase*. If *Washingtonia*'s Pliocene ancestors did grow in a lowland of meandering rivers and riparian streams, which sounds a lot like modern coastal valleys, why don't they grow naturally in modern coastal valleys?

Even if southern California lacked high coastal mountains, rainfall wasn't necessarily ample. A low-lying coast might have been desert if warmer climate pushed the midlatitude arid belt northward. In the 1920s, a geologist likened the Miocene environment of the southern Sierra foothills' richest marine deposit, Sharktooth Hill, to that of today's Topolobampo, a town on the Gulf of California just west of Los Mochis. Sonoran desert grades into tropical thorn scrub there, but the road to Topolobampo seemed venerably desertlike to me while driving it one night, with cardons looming out of the dark and yellow eyes streaking across—a landscape of marble and moon dust.

If California was a tropical land of towering mountains and rain-shadowed basins for most of the Age of Mammals and before, it raises another possibility that seems ironic. Perhaps, instead of evolving in

Mexico and later spreading north, as botanists from Asa Gray to Jerzy Rzedowski have surmised, an "earth-old" desert evolved in California and only spread south into Mexico as global climate cooled in the Pliocene. Anza-Borrego's *Washingtonia* palm fossil wood might be a relic of ancient *California* desert.

All this is pretty thin speculation without more analyses of southern California geohistory. Meanwhile, the new geology's counterintuitive quality makes it hard to even speculate on desert life's past. After all, fossils show that many more kinds of animals lived in southeast Alta California and Baja before the ice age than afterward. It seems that the region should have been more verdant, more arcadian, to support them. Death Valley has yielded thirty-million-year-old fossils of titanotheres, ancient rhinoceros relatives almost as big as elephants. I can't see giant rhinos thriving there today.

Yet rhinos live in southern African deserts today and may have lived in them long enough to shape the vegetation, as Daniel Janzen suggests. Many deserts without much wildlife now supported more even during historical times. A little over two millennia ago, the Greek adventurer Xenophon described what is now western Iraq: "In this region the ground was unbroken plain, as level as the sea, and full of wormwood [*Artemisia*]; and whatever else there was on the plain by way of shrub or reed was always fragrant, like spices; trees there were none, but wild animals of all sorts, vast numbers of wild asses and many ostriches, besides bustards and gazelles."

California never had gazelles, ostriches, or bustards—turkey-size plovers. But it had large herds of equids and pronghorns as well as camels and mastodons for at least twenty million years. The fact that artists conventionally show them living among oaks and cottonwoods instead of creosote bushes and Joshua trees doesn't prove that they always did. It seems that we still just don't know enough about what was happening in North America before two million years ago to be sure when or how the desert evolved. And, given the twenty-first century's information explosion and the consequent fragmentation of scientific disciplines, new paradigms as coherent as Axelrod's may be indefinitely postponed.

The diversity of bushes and lizards that struck me at Mitchell Caverns three decades ago remains a riddle, especially in the case of the big bush-eating lizards. Two lizard experts observed in 2003 that plant eating has evolved several times in the group to which chuckwallas and desert iguanas belong, concluding that "exactly what drove this remains a mystery, but herbivory has one clear benefit: in many environments where temperatures are warm, plants appear to be a virtually unlimited resource. Even in deserts with low productivity, plants are unlimited for short-term periods, and lizards are capable of long-term fasting. So the impetus may simply have been the superabundance of plants relative to lizard abundance."

That is not much of a conclusion in terms of exactly when and how California desert originated. Did lizards start eating desert bushes when megafauna that had preempted lusher flora dwindled? If so, did they start many millions of years ago, maybe so long ago that the dwindling megafauna was dinosaurian? Or did they start a few million years ago when the dwindling megafauna was mammalian? Or did bushes and lizards coevolve in a different way? The Keeper of the Seals still says nothing.

THIRTY-ONE The Falcon and the Shrikes

It is striking if not surprising how quickly the desert empties of humans as the summer nears. When I visited what was then Joshua Tree National Monument in mid-April 1984, the parking lots in the western, high-desert section looked like suburban malls' and the crannies among the nearby granite boulders were full of tents and sleeping bags. When I visited the recently upgraded national park in late May of 1998, the only other person at one large lot in an already blazing dawn was a crazy young Australian in a black raincoat. It got hot enough a few days later that a coyote with fur so short it might have been trimmed at a local pet salon snoozed in a roadside picnic shelter undisturbed by the tourists cruising by in their air-conditioned cars.

On that same afternoon, however, I encountered a scene of strenuous excitement in the Joshua tree forest of Lost Horse Valley. An angry prairie falcon was harassing a pair of loggerhead shrikes, stooping at them as they lurked in the spiky trees, climbing, hovering and shrieking, then stooping again. The shrikes stayed in the foliage but otherwise seemed unconcerned by the falcon's tantrum, hopping around with their usual sluggish inquisitiveness, occasionally emitting hoarse squawks. I hadn't seen a falcon squabbling with shrikes before, although it's probably not that unusual, especially in desert. The shrikes must compete with raptors for sparse prey, and may threaten their nestlings.

From an anthropomorphic viewpoint, a falcon is a straightforward, go-ahead sort of bird. It belongs to a venerable lineage of raptors that, if they live by killing, at least do so in a visibly prompt and efficient way. By contrast shrikes seem creepy and aberrant. Although they are "songbirds" that look more like the inoffensive vireos than hawks, they make a living by impaling lizards and other small prey on thorns and spikes, certainly an ingenious adaptation, but one that Darwin might have included in his Devil's chaplain's book on the "low & horribly cruel works of nature." They usually kill the prey before impaling it, and they impale it because they lack a falcon's talons for holding it while they eat, but the practice still looks nasty.

The agitated falcon see-sawing back and forth over the unflustered shrikes seemed emblematic of civilization's relationship to desert. If they weren't so widely distributed, the raucous black and gray "butcher birds" might be considered desert specialists, and shrikes are especially common in deserts, watching for unwary lizards from ocotillos or other handy thorn bushes. The falcon's behavior reminded me of a pendulum's swings back and forth, and that is how civilization seems to have related to a biome it has found creepy and aberrant.

Like ancient northwest Californians, the cultures of forested western Europe regarded the world as relatively unchanged since distant origins. Greek and Roman literacy gave their myths an air of history, but they didn't emphasize major earthly changes, and thinkers like Aristotle tended to assume that the world was virtually ageless. Cultures of the drier Middle East, on the other hand, resembled those of ancient southern California in seeing the world's origin as something relatively recent that did involve significant change. When they became literate, their mythologies treated stories like the biblical Flood as historical events.

Middle Eastern cultures also resembled southeast California ones in that their myths didn't personify desert, which suggests a similar ambivalence. Early civilizations feared the sandstorms, droughts, and marauding barbarians that regularly emerged from desert, but they also had old traditions for regarding it as a "supermarket." Repeated biblical mention of locusts and manna, foods derived from desert lichens and plant exudates, demonstrate this. And, as the Bible's burning bush shows, desert was an abode of the divine as well as the demonic, although, like the desert, that divine had a scarily unforgiving side. The Middle Eastern Sphinx may have been an aspect of this. The cherubim who guarded the Ark of the Covenant in Solomon's Temple are thought to have been Sphinxes, suggesting that those who guarded the Garden of Eden were too.

If the Middle Eastern Sphinx was a guardian, its Greek transformation into a trickster reflected cultural and environmental shifts as civilizations interacted. Greek and Roman travelers often found deserts frighteningly alien. Herodotus wrote that fox-size, man-eating ants inhabited them and that "immense numbers" of bat-winged snakes flew from Arabia into Egypt each spring. He claimed to have seen heaps of their bones in mountain passes. Meanwhile, the Mediterranean coast was becoming ominously more desertlike from deforestation and overgrazing. As one of Plato's dialogues observed, the fat of the land was wasting away, leaving food only for bees—flowering bushes.

Middle Eastern mythologies' eventual displacement of native ones in Europe perpetuated a fearful sense of desert. As the Bible came to dominate their thinking, Greeks and Romans conflated desert with Christian ideas of the Fall, seeing it as the godforsaken wasteland into which sinful humanity had been banished. Old Testament prophets had gone into the desert to escape the Levantine cities' devilish polytheism. New Testament saints did so to wrestle with the Devil himself. The desert's fiery wastes became emblematic of the torments awaiting the damned at the world's imminent end.

After the Renaissance, when naturalists perceived the great depths of rock strata and inferred that presently observable processes of land uplift and erosion were ancient, Europe swung back toward regarding earthly time as a lengthy continuum with a substantial prehistory. Yet, although they included the whole planet in this conception of "natural" origins, Europeans continued to see deserts in an infernal light. Early in the *Beagle* voyage, Darwin learned to regard the earth's history as an ancient steady state from reading Charles Lyell's *Principles of Geology*. He still shrank from South American desert as a ruined wasteland.

Darwin's confusion reflected the nineteenth-century scientific schism known in geology textbooks as "catastrophism vs. uniformitarianism." Before reading Lyell, young Charles had shared a then common "catastrophist" assumption that, although the earth is old, its formation included episodes of greater change than operate at present—abrupt mountain upheavals, continent-wide floods. Lyell's outlook was called "uniformitarian" because he doubted that the earth's past included episodes so different from its present.

According to textbook convention, the schism pitted a religious view that sudden supernatural change formed the earth against a scientific one that gradual natural processes have done so, with scientific evidence prevailing in the end. Historians have shown that this is oversimplified. "Catastrophist" scientists like Louis Agassiz based their belief in sudden change as much on natural evidence as on biblical traditions. "Uniformitarian" ones like Lyell based their belief in gradual change as much on philosophical traditions going back to Aristotle as natural evidence. Indeed, the schism persists in secular terms today as exponents of sudden wholesale mass extinctions by comet impact argue with exponents of gradual piecemeal mass extinctions by climate change. So, although Lyell's "steady state" vision convinced Darwin intellectually, his overall attitude to desert continued to vacillate between the poles of sudden and gradual change, and later attitudes to desert did too.

Most early California naturalists were catastrophist in sharing Darwin's sense that desert is a ghastly wasteland caused by a relatively recent and sudden episode of mountain formation. Some swung toward uniformitarianism in sharing his post-*Origin of Species* sense that aridity has restricted desert life's development to a few slowly evolving, presumably very old forms. Later some swung back toward catastrophism by proposing that aridity might stimulate evolutionary change instead of obstructing it. Although anti-Darwinians originated such ideas, neo-Darwinians eventually incorporated them into their synthesis of genetics and natural selection. G. Ledyard Stebbins's ideas of a shift to accelerated "quantum evolution" as California's climate grew drier supported Daniel Axelrod's of a new "regional desert" created by geologically recent changes so unusually severe as to seem catastrophic.

Despite public indifference to neo-Darwinism, California civilization seems to have assimilated a "quantum" sense of the desert as a dynamic, invasive phenomenon. John Steinbeck presented it as a lurking threat to coastal Arcady in his novel, *To a God Unknown,* wherein drought strikes northwestward like a monstrous claw to crush a homesteader: "He rode slowly home by the bank of the dead river. The dusty trees, ragged from the sun's flaying, cast very little shade on the ground. . . . The hills were gaunt now; here was a colony from the southern desert come to try out the land for a future spreading of the desert's empire. . . . A horned toad [lizard] came out of the dust and waddled to the bottom step of the porch, and settled to catch flies."

A newspaper description of Death Valley contemporaneous with Steinbeck's novel reads like Herodotus on ancient Arabia: "In the early spring gnats of great size sweep over the place in swarms of millions. Coyotes prowling around the fringes of the valley have been stung to death. . . . There is a peculiar kind of mouse which lives wholly on the scorpions which thrive there by the tens of thousands. Another variety of mouse consumes the centipedes." This sense of threat multiplied in the megalopolis that mushroomed from the homesteaders' Arcady.

One neo-Darwinist aspect that did enter public consciousness with a vengeance was the role of genetics in evolution. Hollywood was quick to exploit this as the reanimated corpses of 1930s horror movies morphed into the genetically altered monsters of 1950s ones. I have a nightmarish childhood memory of the 1954 movie *Them!,* which echoes Herodotus with man-eating desert ants mutated to automobile size by atomic radiation. Spawned in a bomb test site, the winged queens quickly colonize the Los Angeles sewer system. This is complete fantasy—insect physiol-

ogy restricts them to little more than mouse size regardless of mutation. Still, I once talked with someone who'd visited nuclear test sites
in Nevada and thought "lower animals" did better in the radioactive
ambience than birds and mammals, "as though evolution was going
backward."

Such fantasies and anxieties implicate a mutation-haunted, invasive
desert as an impending retribution for civilized excess. As one chroni-
cler of the genuinely murderous desert "freaks" known as the Manson
family wrote:

> There is something uneasy in the Los Angeles air this afternoon, some
> unnatural stillness, some tension. What it means is that tonight a Santa
> Ana wind will blow, a hot wind from the northeast whining down
> through the Cajon and San Gregorio passes, blowing up sandstorms
> along route 66, drying the hills and nerves to the flash point. . . . I am
> not pleased to see, this year, cactus spreading wild to the sea. There
> will be days this winter when the humidity will drop to ten, seven,
> four. Tumbleweed will blow against my house and the sound of the
> rattlesnake will be duplicated a hundred times a day by dried bougain-
> villeas drifting up my driveway. . . . Los Angeles weather is the weather
> of catastrophe, of apocalypse.

An irony of this lurid picture is that neither of its scientific progenitors
saw the desert so. Stebbins and Axelrod both shared Edmund Jaeger and
Joseph Wood Krutch's sense of desert as a refuge from an increasingly
threatening civilization. This sense may have nudged Stebbins toward
regarding it as an evolutionary museum instead of an evolutionary frontier. He helped to start the California Native Plant Society, one of the
state's main desert preservation advocacies. Axelrod, with characteristic
ingenuity, interpreted a recent desert as an especially vulnerable biotic
treasure: "Our regional deserts represent *new ecosystems:* they have just
been born! Unless we take better care of them, nothing will remain
but a barren terrain like the largely *man-made* desert that now stretches
uninterruptedly for 4,200 miles, from the Atlantic shore of North Africa
to the Thar desert of western India. The choice is yours."

Now Stebbins's museum desert seems to be prevailing over Axelrod's
newborn one. The pendulum is swinging back toward uniformitarian-

ism as geologists and paleobotanists begin to perceive western mountains and possibly deserts as much older than most early naturalists had imagined. This tells us little more than we already know about the evolution of desert plants and animals, but it does imply that there has been as much evolution of desert plants and animals as of forest or grassland ones.

It is hard to see how a California civilization that is used to rapid change will respond to this. Tractmongers don't care if the bushes they bulldoze have been here three million years or thirty million. Even some prominent desert admirers seem indifferent to its evolutionary past. One of the late twentieth century's best books on the subject is *Scenes in America Deserta*, which echoes John Van Dyke's *The Desert* in that its author, Peter Reyner Banham, was an art history professor (at U.C. Santa Cruz) who reveled in the arid landscape's spectacular colors and atmospheric displays. "I had been granted a vision of the desert's ultimate splendor," Banham writes of a "stratum of luminous white mist" seen above a dry lakebed in the Mojave. "My sensibility was transfixed, my consciousness transformed." Unlike Van Dyke, however, Banham shows little interest in desert origins, either geological or evolutionary. Fascinated by its often bizarre interactions with American technological culture, he tends to regard desert as a figment of the human imagination:

> The cliché "man-made desert" has become so worn with use that we hardly hear it any more, but it could be one of the great, awful, boring truisms at the basis of many of our desert attitudes. The self-conscious loner who goes out into the desert to "find himself" may do so in more than just the narcissistic sense the phrase normally intends—he may still be staring into the face of man. . . . Ultimately deserts are man-made in what may be a culturally important sense . . . all deserts are deserts *by* definition . . . *desert* is a concept of, and about, people. . . . The ultimate definition of a true desert may yet prove to be concerned with the number and type of people present, and what they think they are doing there.

Banham is right psychologically. There were no desert lovers until the term was invented and applied to arid places. Miguel del Barco, despite his fascination with Baja California's weird plants, was certainly not one. Still, Banham's disregard for desert deep time can lead him astray, as when he tries to explain the name of the place east of Death Valley where I watched Amargosa pupfish in their deep blue spring holes. Banham assumes it is a cultural reference, albeit one he finds puzzling. He regrets his inability

> to hear with any inner understanding the depth of poetry and trag-
> edy implied in a place name like "Ash Meadows." One is apt to be so
> impressed by its blunt functionalism that one tends to lump it with all
> the Bitter Creeks and Sweetwaters one knows. Yet the concept of meadows
> made of ashes is as terrifying as it is poetic—it is a concept that a Ray
> Bradbury might have invented to describe the aftermath of an atomic
> explosion; but as a place name it goes back before both Bradbury and
> the Bomb (when it was a tent city staging post and the site of a short
> lived clay pit operation) and presumably represents some profound
> disappointment on the part of an early voyager. . . . But could it mean
> simply meadows with ash trees? I doubt it. Not in this landscape.

In fact, that is exactly what it means. Ash Meadows is named for the scattered, relict groves of stunted leatherleaf ash trees (*Fraxinus velutina*) that the springs' fossil groundwater has allowed to survive there since climate dried after the last ice age. For an "inner understanding" of the place, Banham needed to go back much farther than he thought. Originally from England with its traditions of Arabian exploration, his first response to the Mojave was to wonder where the camels were. He might well have been standing on the buried bones of fossil ones.

Of course, better public awareness of the desert's deep past won't extricate California civilization from the troubles that its excesses have brought upon it. New evidence of ancient stability won't allay suburban panic as desert winds blow stronger and wildfires burn longer with every year of "global warming." It might add credence to the sense— expressed by writers like Jaeger and Krutch, and by parks and wilder- ness areas—that the desert is not a backward, ruined wasteland but an ingenious, healthy response to environmental constraints primal and inevitable on this planet.

For all his shrinking from New World desert, Darwin seems to have felt something of this in retrospect. *The Voyage of the Beagle* ends on a note of puzzled nostalgia:

In calling up images of the past, I find the plains of Patagonia frequently cross before my eyes; yet these plains are pronounced by all wretched and useless. They can be described only by negative characters; without habitations, without water, without trees, without mountains, they support merely a few dwarf plants. Why then, and the case is not peculiar to myself, have these arid wastes taken so firm a hold of my memory? Why have not the still more level, the greener and more fertile Pampas, which are serviceable to mankind, produced an equal impression? I can scarcely analyze these feelings: but it must be partly owing to the free scope given to the imagination. The plains of Patagonia are boundless, for they are scarcely passable, and hence unknown: they bear the stamp of having lasted, as they are now, for ages, and there appears to be no limit to their duration during future time.

Epilogue

I didn't go back to Red Rock Canyon for years after 1983. My first impression of ominous squalor morphing into surreal enchantment seemed unrepeatable. And the canyon was different when I did return in spring 2008—a tidily improved state park on a rebuilt, four-lane highway. Litter and ORV tracks were scarce. Ubiquitous signs protected raptor-nesting cliffs and botanical features. Well-maintained trails led into the formations.

But surrealism lingered in the hoodoos and minarets. I soon found myself walking a narrow gulch strangely unmarked by recent human presence. There was broken glass, inescapable in a former movie location, but even it seemed ancient. Winter rains had washed the twinkling shards into the ten-million-year-old Miocene strata. The only tracks in the sand were a small fox's.

The gulch led to a cliff of yellow tuff eroded to the semblance of an Egyptian necropolis. (A similar cliff looms above the unearthing of a royal tomb, probably filmed at Red Rock, in the Karloff version of *The Mummy*.) At the top was a stratum of sarcophagus-like formations; below it, mummylike ones; below that, forms resembling the bulbous, elongate crown of Upper Egypt. At the bottom, paw-shaped formations raked the sand into cavelike recesses. An oblong monolith thrust from the cliff's center like a featureless face. Behind the yellow cliff, a taller red one hinted at further fantasies.

If I was to meet the Sphinx, I thought, this would be the place. But what would the California Sphinx be? Maybe a mountain lion with condor's wings and—given the Hollywood connection, with apologies to Mary Austin—the face and bust of the scariest screen goddess, Bette Davis. The Greek Sphinx was said to be gorgeous as well as sexually predatory. Sophocles called her "the cruel enchantress." And what riddle would the California Sphinx pose? Joseph Smeaton Chase said there is no answer to her riddle; he didn't say what the riddle is.

But the Keeper of the Seals said nothing at the yellow cliff. I found not even a wood rat den in its recesses, much less any figment of Bette Davis's fabled bust. As the midday sunlight crept down the ranks of Egyptian phantasms, the broken bottles' glitter started to give me a headache. I retreated to the cinderblock visitor center where a cactus wren wheezing in the shrubbery and a mockingbird squawking on the flagpole implied a return to premigraine normality.

As my eyes adjusted to the fluorescent lights, however, I saw that the riddler had been waiting for me there in air-conditioned comfort. The center's little museum was a monster's lair of bizarre bones dug from the canyon—bones of elephant-like gomphotheres, doglike borophagids, lionlike nimravids, and oreodonts, extinct ungulates unlike any living beast. The first bones I saw were those of a Miocene camel, *Megatylopus*, found in 1998. It stood sixteen feet tall and had weirdly elongated neck vertebrae like a giraffe's. Displayed nearby were the fossil ankles of five camel species ranging from deer to giraffe size.

There were plant remains too, from a park deposit that had yielded oak, pine, acacia, locust, palm, and cypress fossils. The display said the area had been "lush savanna" in the Miocene. But it also said the deposit had included two of the desert bushes I first encountered in the Providence Mountains—the tomato relative *Lycium* and the rose relative blackbush (*Coleogyne*).

The lair's collection was predictably ambiguous. On the surface, it displayed the Madro-Tertiary paradigm with its woodlands and savannas, a region of low hills with a few volcanoes or granite outcrops providing semiarid habitat for *Lycium* and blackbush. Axelrod's last desert fossil publication describes this from Miocene epoch material he found nearby: "The small Ricardo flora (12 m.y.) from the El Paso Mountains, known chiefly from fossil wood, includes a dominant oak associated with pinon, cypress, palm (possibly *Sabal*), a legume (*Robinia* [black locust]?), buckbrush (*Ceanothus*), and a few leaves representing dry tropic scrub taxa (*Acacia, Lycium*). The sample indicates that there were semidesert patches in the nearby area, and that rainfall over the region probably was not less than 18 to 20 inches. . . . The modern Mohave desert as we see it now is certainly younger than 10,000 years."

But behind this display lurked the outlines of a Miocene California more like the present than seemed possible a few years ago, a land of towering forested ranges and rain-shadowed basins. If the analyses of the past two decades are right, such stability may be more natural than we imagined, although the thought of so little change through tens of millions of years is less familiar than the extinct monsters whose bones rest in the park's little museum. Yet Red Rock fossils also show, as at Anza-Borrego, that animals very like today's lived with the extinct monsters—"desert" reptiles like spiny lizards and night lizards; "nondesert" ones like alligator lizards.

To be sure, the outlines of that unfamiliar past were faint. Little is known about Red Rock Canyon plant fossils because dealers cleaned out most of the deposit before the park's establishment in 1968. When I asked the attendant if any plant fossils had turned up lately, she showed me a rock with ferny patterns on it. "What does this look like?" she asked. I cautiously equivocated that it looked as much crystalline as vegetative. She nodded, "I was excited when I found it, but when I showed it to a geologist, he said it was 'dendrites.' Oh well, it's pretty."

I sidled back to the display and found another of the Sphinx's tro-
phies—the skin of a small black bear killed by a truck on the highway
some years before. It must have wandered down from the Sierra forests
to the west, but its presence in Red Rock Canyon still seemed bizarre. I
wondered if its bones had gone into a landfill along with Christmas trees
and garden clippings, perhaps to suggest to some hard-to-imagine paleontologist thirty million years from now that forests covered southeast
California in the Age of Man.

With that thought, I plunged back into the midday sun, vaguely want-
ing some kind of closure to desert ambiguities. I wandered to the Red
Cliffs Botanical Area, which, although overlooked by hoodoos, is one of
the park's less surreal scenes—a sandy ridge by a wash with the El Paso
Mountains' scaly gray slopes on the other side. There, bushes and lizards
reigned, resplendent in spring reproductive finery.

I watched a side-blotched lizard for a while. With the naked eye,
its coloring was indistinguishable from the sand; with binoculars, it
resolved into a gay pattern of orange dots on azure. Nearby was one
of the more spectacularly flowering bushes I've seen, although only
two feet tall—a spherical mass of white flowers so fragrant that it
seemed to have as many violently buzzing bees and flies as blossoms.
The four-petal flowers made me think it was a mustard relative, but I
wasn't sure. A tiger whiptail ran past, also looking sand colored until
I focused the glasses on its yellowish-bluish-purplish checker pattern.
It disappeared under another spherical, bee-filled bush with lobed
lavender flowers. I thought it might be a mint relative, but I hadn't seen
one before.

I was glad to find some bushes I recognized, a rough circle of rangy,
olive drab ones. It was a creosote bush clone, a genetic individual that
had begun life as a normal vegetable Don Quixote but had outlived its
generation. As its original stems died, vegetative offshoots had spread
out in a desert fairy ring. Botanists have estimated, judging from the
species' slow growth, that large clones may have lived over ten thousand
years—since the last ice age. This one was not big enough for that, but
a few thousand years seemed possible. Maybe it was older than a big
redwood, although redwoods also clone so that a two-thousand-year-old

tree may be a much older genetic individual. Creosote bush clones seem less venerable, but if the desert has been in California since the Miocene or before, who knows? A clone ring might die, but part of it might survive and clone again.

The clone offered shade only for lizards, so I moved down-ridge into a patch of sand hummocks peppered with small Joshua trees. One clump might have been a bonsai oasis, except that the nearest hint of water was an alkaline smudge down in the wash. I crawled into its thin shade anyway, and dozed awhile. Waking, I noticed a circular sandy depression wherein grew stunted creosote bush, burroweed, cholla, and flowers of goldfields, purple mat, and desert chickory. It was oddly regular in shape, and when I looked closer, I saw that it had a concrete rim.

Who would build a fifteen-foot wading pool in such a place? A park map suggested an answer. It showed a dotted circle labeled "Historic Movie Set: *Water Hole No. 3* (dry)." Taking this literally, I imagined Model-Ts baking in the wash while a jodhpured hack brandished a megaphone beside the pool. Some bottle shards seemed of that vintage. A sun-bleached green one read, with antique candor: "Carbonated Citrus Beverage." But the answer to the wading pool riddle—found on a DVD—was stranger. *Water Hole No. 3* is no quaint silent or B-talkie: it is widescreen Technicolor, made in 1967. And it is not a typical oater with a hero who vanquishes the outlaws and rescues the maiden but a Vietnam-era anti-Western with an antihero who rapes the maiden and abandons her to the outlaws. It flaunts the excess for which Californians fear desert retribution.

The art director might have mined *The Land of Little Rain* for atmosphere, because the movie echoes its ticky-tack mining towns spilled out of gulches. "Things had a way of happening at Red Rock," wrote Mary Austin, who often passed on the stagecoach to Owens Valley and in her autobiography called it "one of the weirdest wind-sculptured defiles of the West." A famous passage in *The Land of Little Rain*'s first chapter might have inspired the movie's plot: "The palpable sense of mystery in the desert air breeds fables, chiefly of lost treasure. Somewhere within its stark borders, if one believes reports, is a hill strewn with nuggets; one seamed with virgin silver; an old clayey water bed where Indians scooped up earth to make cooking pots and shaped them reeking with grains of pure gold."

The inspiration didn't go far. The film's scenario of army gold stolen by outlaws and hidden in the wading pool evokes little mystery. The gambler antihero, played by squint-eyed James Coburn, steals a map to the treasure via "Water Holes 1 and 2" and rides blithely off as if oases are really so handy. A scene wherein he wallows after the gold in the pool might be lifted from Clarence King's oasis idyll, but it evokes no sense of romantic discovery. The film doesn't even make much of the canyon's geological grotesqueries, relegating them to red and yellow blobs in the background.

Far from bringing closure to desert ambiguities, the wading pool presented another one. *Water Hole No. 3* is among 130 films made in Red Rock Canyon, and such industrial disturbances cause endless arguments. Environmentalists say desert is so hostile that its life is fragile, citing persistent tank tracks in General George Patton's World War II California Desert Training Center. Antienvironmentalists say desert is so hostile that its life is resilient, citing vanished tank tracks in Patton's DTC.

Red Rock Canyon hasn't undergone army maneuvers but it still is one of the desert's more disturbed places. Joseph Chase was already complaining of "those nuisances the motion-picture people" in 1919, and later growth brought others. ORVs scraped twelve thousand tons of topsoil from just one mile-long hillside there. The state park is an improvement, but the highway obviates any return to pristine conditions. Species like tortoises suffer and they, not tank tracks, are the measure of desert fragility. When I arrived via Tehachapi Pass in 2008, my "quixotic" perception of creosote bush seemed prescient. The bushes around the pass tilt at windmills—the "green energy" turbines whose blades kill the avifauna the state park tries to protect.

I don't know what significance the Chinese box of desert ambiguity has for a civilization that, despite all its "sustainability" talk, remains intent on straightforward growth. Most Californians think technology is the answer to the desert's challenges. Some believe it can "transform" the desert with desalinated seawater and limitless new energy sources.

Some environmentalists mourn "the end of nature" in its Egyptian Sphinx-like role as a guardian of terrestrial order. Because none of the biosphere remains uninfluenced by humanity, they conclude, technology has replaced nature. John Van Dyke expressed similar sentiments in sepia prose: "What monstrous folly, think you, led Nature to create her one great enemy—man! . . . She learned the meaning of fear when this new Oedipus of her brood was brought forth!"

Maybe the California Sphinx will self-destruct if "this new Oedipus" answers her riddle. But, again, what would the riddle be? Maybe it would be the same as the Greek Sphinx posed to the old Oedipus: "What goes on four legs at dawn, two at noon, and three at dusk?" Evolution has raised new implications for that. One might say that humanity went on an ape's four legs in a Miocene dawn, on a hominid's two legs in a Pliocene and Pleistocene noon, and on a technologist's three legs after *Homo sapiens* prevailed. The crutch must have been one of the earliest tools.

Of course, the old Oedipus story implies second thoughts about answering mythic monsters' riddles. Sophocles' hero brags of outsmarting the Sphinx after the blind, androgynous prophet Tiresias refused to try: "I came along, yes I, Oedipus the unlearned, and stopped her by using thought, not augury of birds." But he has no more eyes for the Sphinx's identity than for his true parents'. He sees only obstacles to his progress, and he progresses into three-legged twilight.

Nature the guardian may be finished, but nature the trickster is not. High technology can add solar collectors, wind farms, and CO_2 scrubbers to the mines, agribusinesses, highways, and power plants that seem to trump the desert. It can't remove earthquakes, wildfires, and storms. "Current attempts and plans to further 'develop' and 'improve' these arid landscapes represent the dreams of the largely uninformed," wrote Daniel Axelrod. "Man has not yet improved on nature. He never will." Controlling nature, to the limited extent that it is possible, depends on self-control. Oedipus's fate reflected his lack of it.

Away from the highway, anyway, desert still trumped technology at Red Rock Canyon in 2008. Besides broken glass, bits of machinery so rusty they looked melted were all that remained of ticky-tack towns, either the real Old West's or Hollywood's. Of course, the concrete wading pool remained, and it may last for centuries in the dry sand. Yet the movie that left it behind like a bottle by the road has an odd way of evoking desert permanence.

For all its noisy banality, a surreal element emerges from *Water Hole No. 3*'s increasingly chaotic story, especially when the down-market DVD sticks and reduces the action to a lurching mosaic of blobs and squawks. It begins to seem less a drama than a dream wherein events transpire with the subconscious mind's caprices. Although pushed into the background, the supremely arid Mojave covertly enchants a production that would have been humdrum in another setting. Westerns are supposed to be about wide-open spaces, but the landscape of *Water Hole No. 3* appears less expansive than pensive, as though absent-mindedly tolerating the racket while it pursues immemorial ruminations. This implies a deep imperviousness.

The outlaws gallop past hoodoos, through salt flats, past hoodoos, over sand dunes, past hoodoos. The U.S. Cavalry arrives, intermittently, inexplicably. Bette Davis does not appear, but Joan Blondell, the singing, dancing star of Busby Berkeley's *Gold Diggers of 1933*, does briefly as the madam of an improbably posh brothel, looking as gaudy as the little desert bushes. As the film gets louder and faster—more galloping, shooting, fondling, fisticuffs—the action seems propelled toward a point where it will start to blur and fade, thrust into nonentity by its own manic improbability, leaving only hoodoos, salt flats, sand dunes, and bushes.

> The Worldly Hope men set their Hearts upon
> Turns ashes—or it prospers; and anon,
> Like Snow upon the Desert's dusty Face,
> Lighting a little Hour or two—is gone.
> They say the Lion and the Lizard keep
> The Courts where Jamshyd gloried and drank deep:
> And Bahram, that great Hunter—the Wild Ass
> Stamps o'er his Head, and he lies fast asleep.
> Edward FitzGerald, *Rubaiyat of Omar Khayyam*

Notes

Page

 xi "Go as far as you dare": Austin, *Land of Little Rain*, in *Stories from the Country of Lost Borders*, p. 13.

 xii "Death and life usually appear": Abbey, *Cactus Country*, p. 23.

 xv "The best way of inducing these novices": Cowles, *Desert Journal*, p. 76.

xviii "The more I looked": Shrub genus names: ephedra is *Ephedra;* blackbush is *Coleogyne;* bitterbrush (also called antelope brush) is *Purshia;* indigo bush is *Psorothamnus;* honey mesquite is *Prosopis;* catclaw acacia is *Acacia;* cheesebush is *Hymenoclea;* desert fir is *Peucephyllum;* felt thorn is *Tetradymia;* brittlebush is *Encelia;* desert holly, saltbush, and shadscale are *Atriplex;* blue sage and Mojave sage are *Salvia;* paper-bag bush is *Salazaria;* banana yucca and Mojave yucca are *Yucca.*

 xxi "Surely this is North America's most barren desert": Abbey, *Journey Home*, p. 74.

CHAPTER ONE. A SPHINX IN ARCADY

2   "the fertilizing waters of the rivers": Muir, *Mountains of California*, p. 246.

2   "The mountains, the sea": Chase, *California Desert Trails*, p. 358.

2   "fixed in eternal reverie": Ibid., p. 1.

2   "According to mythologist Joseph Campbell": Campbell, *Masks of God*, p. 91.

CHAPTER TWO. THE COUNTRY OF DRIED SKIN

7   "The age of the artifacts"

7   "began with the assumption": Heizer, *Handbook of North American Indians*, vol. 8, p. 27.

7   "According to George Laird of the Chemehuevi tribe": Laird, *Chemehuevis*, p. 149.

7   "According to anthropologists": Heizer, *Handbook of North American Indians*, p. 658.

7   "According to Francisco Patencio": Patencio, *Stories and Legends of the Palm Springs Indians*, p. 19.

8   "Desert is the name it wears": Austin, *Land of Little Rain*, in *Stories from the Country of Lost Borders*, p. 9.

8   "According to Laird, their word for desert": Laird, *Chemehuevis*, p. 87.

9   "To a white man, the desert is a wasteland": Ibid, pp. 4–5.

9   "there was 'no help'": Austin, *Land of Little Rain*, in *Stories from the Country of Lost Borders*, p. 12.

9   "dark and terrifying place": Laird, *Chemehuevis*, p. 5.

9   "all pervasive and intense feeling": Heizer, *Handbook of North American Indians*, vol. 8, p. 582.

10   "which had the appearance": Manly, *Death Valley in '49*, p. 85.

CHAPTER THREE. A CACTUS HERESY

11   "It was decided": Font, *Anza Expedition of 1775–1776*, p. 117.

12   "Father Garces is so well fitted": Lee, *Great California Deserts*, p. 50.

12   "Since the land is so elongated": Barco, *Historia Natural y Cronica de la Antigua California*, p. 3.

12   "In the great arroyos": Ibid., p. 70.

13   "Even though this tree is permeated": Ibid., p. 84.

14   "The author of nature": Ibid., p. 87.

15   "For the rest . . . the Indians can make": Ibid., p. 216.

CHAPTER FOUR. THE CREATOR'S DUMPING GROUND

16 "a west course fifteen days": Morgan, *Jedediah Smith*, pp. 200, 242.
17 "We traveled 'til long after dark": Fremont and Emory, *Notes of Travel in California*, p. 50.
17 "In no part of this vast tract": Ibid., p. 55.
17 "Crossing a low Sierra": Fremont, *Report of the Exploring Expedition to the Rocky Mountains*, p. 256.
18 "The whole idea": Ibid., p. 277.
18 "Travelers through countries": Ibid., p. 266.
18 "Many of these Indians": Ibid., p. 267.
18 "The country had now assumed": Ibid., p. 262.
18 "twenty miles to the southward": Ibid., p. 257.
19 "As I reached the lower part": Manly, *Death Valley in '49*, p. 86.
19 "One fellow said": Ibid., p. 87.
20 "If the waves of the sea could flow": Ibid., p. 163.
20 "There before us was a beautiful meadow": Ibid., p. 108.

CHAPTER FIVE. AN EVOLUTIONARY BACKWATER

22 "I am tired of repeating": Darwin, *Voyage of the Beagle*, p. 335.
22 "While traveling through these deserts": Ibid., p. 333.
22 "utterly irreclaimable and desert": Ibid., p. 319.
22 "It has been inferred": Ibid., p. 44.
23 "In the spring": Ibid., p. 333.
23 "We must admit": Ibid., p. 375.
23 "To be sure, competition": Darwin, *Origin of Species*, p. 61.
24 "struggle for life": Ibid., p. 56.
24 "Two canine animals": Ibid., p. 52.
24 "In regard to birds": Darwin, *Descent of Man*, p. 809.

CHAPTER SIX. ANTI-DARWINIAN LACERTILIANS

26 "Nothing could be less inviting": Darwin, *Voyage of the Beagle*, p. 359.
28 "the peculiar elements": Gray, "Characteristics of the North American Flora," p. 340.
28 "competition from the Mexican Plateau vegetation": Ibid., p. 326.
28 "He bases his theory": Henslow, *Origin of Plant Structures*, p. 8.
29 "We thus begin to suspect": Ibid., p. 34.

29 "As one anti-Darwinian": Cope, *On the Method of the Creation of Organic Types.*

30 "Darwinism's leading American foe": Lurie, *Louis Agassiz,* p. 376.

31 "Owing to the continued exploration": Gilmore, *Fossil Lizards of North America,* p. 3.

32 "It was very hot": Osborn, *Cope: Master Naturalist,* p. 262.

32 "As Professor Marsh does not give us any clue": Cope, "Reptiles of the American Eocene," p. 981.

32 "All of the fossil lizard remains": Gilmore, *Fossil Lizards of North America,* p. 2.

CHAPTER SEVEN. DESCRIPTIVE CONFUSION

35 "level floor, as white as marble": King, *Mountaineering in the Sierra Nevada,* p. 23.

35 "Sheets of lava": Ibid., p. 34.

35 "Spread out below us": Ibid., p. 41.

35 "The California deserts": Brewer, *Up and Down California,* p. 535.

36 "I found the so-called desert": Muir, *Mountains of California,* p. 73.

36 "It is here that the desert": Coville, *Botany of the Death Valley Expedition,* p. 8.

36 "It is in the shrubby vegetation": Ibid., p. 43.

36 "the excessive dryness": Ibid., p. 5.

37 "The vegetation has the characteristic": Jepson, *Manual of the Flowering Plants of California,* p. 4.

37 "their relationships and origins": Ibid., p. 10.

37 "irresistible fascination": Beidleman, "Willis Lynn Jepson," p. 285.

37 "furnish the breeding spot": Beidleman, "Rowboat Botanizing with Willis Linn Jepson on the Colorado River, 1912," p. 5.

38 "The scantiness of the desert vegetation": Coville, *Botany of the Death Valley Expedition,* p. 43.

38 "the report compiled by the expedition's leading zoologist": C. H. Merriam et al., *Death Valley Expedition,* p. 171.

38 "the ferocity and greed": Ibid., p. 167.

39 *"Bythinella protea"*: Ibid., p. 278.

39 "supersaturated with the bitter chemicals": Ibid., p. 180.

41 "a palpable sense of mystery": Austin, *Land of Little Rain,* in *Stories from the Country of Lost Borders,* p. 16.

41 "If the desert were a woman": Austin, *Lost Borders,* in ibid., p. 160.

41 "It is recorded in the report": Austin, *Land of Little Rain,* in ibid., p. 11.

41 "There is neither poverty of soil": Ibid., p. 13.

42 "So wide is the range": Ibid., p. 38.

42 "Watch a coyote," Ibid., p. 23.

42 "If one is inclined to wonder": Ibid., p. 15.

42 "It is the opinion of many": Ibid., p. 22.

42 "There are myriads of lizards": Ibid., p. 89.

43 "herb-eating, bony-cased old tortoise": Ibid., p. 65.

43 "Van Dyke's firsthand knowledge": Wild, "Sentimentalism in the American Southwest," p. 133.

43 "The afternoon sun": J.C. Van Dyke, *Desert,* p. 229.

44 "It is a gaunt land": Ibid., p. 26.

44 "Everywhere you meet with the dry lake-bed": Ibid., p. 35.

44 "Nature goes calmly on": Ibid., p. 62.

44 "tender snow-flowers": Muir, *Mountains of California,* pp. 22–23.

45 "I need not now argue beauty": J.C. Van Dyke, *Desert,* pp. 191–192.

45 "And always here in the desert": Ibid., p. 130.

45 "The life of the desert": Ibid., pp. 150, 171.

45 "The deserts should never be reclaimed": Ibid., p. 59

45 "But, with his lung problems": Ibid., p. xlv.

46 "wrangled incessantly": D. Van Dyke, *Daggett,* p. 135.

46 "Muir retreated 'in great disgust'": Ibid., p. 106.

46 "The desert animals": J.C. Van Dyke, *Desert,* p. 151.

46 "It would seem as though Nature": Ibid., p. 167.

47 "It is as if you were bemused": Chase, *California Desert Trails,* p. 2.

47 "One feature of loveliness": Ibid., p. 4.

47 "I do not see how Sahara": Ibid., p. 275.

47 "With ingenious pains": Ibid., p. 53.

48 "With all my weariness": Ibid., p. 280.

48 "bony little goblins": Ibid., p. 89.

48 "To the astonishment of everybody": Lee, *Great California Deserts,* p. 223.

48 "There are, it is true": Chase, *California Desert Trails,* p. 4.

49 "It has been swept by seas": J.C. Van Dyke, *Desert,* p. 229.

49 "What a book a Devil's chaplain": Darwin, *Correspondence of Charles Darwin,* vol. 6, p. 178.

CHAPTER NINE. HOPEFUL MONSTERS

51   "Desert conditions": MacDougal, "Influence of Aridity upon the Evolution-
     ary Development of Plants," p. 231.
51   "The view that such forms": Ibid., p. 227.
51   "Adaptation, therefore, furnishes": MacDougal, "Origin of Desert Floras,"
     p. 118.
52   "Gradual modifications": MacDougal, *Botanical Features of North American
     Deserts*, p. 106.
52   "It is true, of course, that desert conditions": MacDougal, "Influence of
     Aridity upon the Evolutionary Development of Plants," p. 227.
53   "With the exception of the John Day region": J. C. Merriam, "Extinct Faunas
     of the Mohave Desert," p. 245.
53   "In a few strata abundant remains": Ibid., p. 251.
53   "As nearly as the writer can judge": Ibid.

CHAPTER TEN. AN OLD EARTH-FEATURE

55   "One of the most commonly held ideas": Shreve, *Cactus and Its Home*, p. 29.
56   "There are strong reasons": Sumner, "Some Biological Problems of Our
     Southwestern Deserts," p. 370.
56   "Great stress": Ibid., p. 357.
56   "Despite arguments": Ibid., p. 370.
57   "The question before us": C. H. Merriam, "Is Mutation a Factor in the Evo-
     lution of Higher Vertebrates?" p. 243.
57   "What does this mean?": Ibid., p. 246.
57   "He interpreted creosote bush": Clements, "Origin of the Desert Climax
     and Climate," p. 122.
58   "There is evidence": Shreve and Wiggins, *Vegetation of the Sonoran Desert*,
     vol. 1, p. vi.
58   "His field work in Chile": Howard, "Ivan Murray Johnston, 1898–1960,"
     p. 2.
59   "clearly a South American type": Johnston, "Floristic Significance of Shrubs
     Common to the North and South American Deserts," p. 357.
59   "When it is realized": Ibid., p. 360.
60   "Since most biologists": Ibid., p. 361.
60   "Johnston cited examples": Ibid., p. 362.

CHAPTER ELEVEN. A CLIMATIC ACCIDENT

61 "A 1944 textbook": Cain, *Foundations of Plant Geography*, p. 121.
62 "In my junior year": Axelrod, "Response of Daniel I. Axelrod for the Award of the Paleontological Society Medal," p. 522.
62 "Fossil evidence bearing directly": Cain, *Foundations of Plant Geography*, p. 119.
63 "The occurrence of a desert element": Axelrod, "Pliocene Flora from the Eden Beds," p. 4.
63 "True desert conditions": Axelrod, "Pliocene Flora from the Mount Eden Beds, Southern California," p. 139.
64 "A site from the Eocene": Axelrod, "Miocene Flora from the Western Border of the Mohave Desert," p. 61.
64 "A wide diversity of opinion": Axelrod, "Evolution of Desert Vegetation in Western North America," p. 217.
65 "Unconcerned with ideas of plant succession": Barbour, "Ecological Fragmentation in the Fifties," p. 239.
65 "Clements's postulate": Axelrod, "Evolution of Desert Vegetation in Western North America," p. 219.
65 "There appears to be no support": Ibid., pp. 289, 292.
66 "Such migrations account for": Ibid., p. 298.
66 "Increasing topographic and climatic diversity": Ibid., p. 293.
67 "The task of determining the origins": Ibid., p. 288.
67 "The *Larrea-Franseria* desert": Cain, *Foundations of Plant Geography*, p. 121.
68 "I thought by all means": Mead, "Life and Work of G. Ledyard Stebbins," p. 48.

CHAPTER TWELVE. AN EVOLUTIONARY FRONTIER

71 "Rapid evolution requires": Stebbins, "Evidence of Rates of Evolution from the Distribution of Existing and Fossil Plant Species," p. 156.
72 "The book, which synthesized perspectives": Crawford and Smocovitis, *Scientific Papers of G. Ledyard Stebbins*, pp. 7, 21.
72 "The new combined attack": Stebbins, "Aridity as a Stimulus to Plant Evolution," p. 33.
73 "In the first place, where moisture": Ibid., p. 35.

CHAPTER THIRTEEN. A NEO-DARWINIAN GALAPAGOS

76 "One night we had a fair camp": Manly, *Death Valley in '49*, p. 82.
78 "When a biologist stands": Deevey, "Biogeography of the Pleistocene," p. 1397.
78 "Just as biological processes": Soltz and Naiman, *Natural History of Native Fishes in the Death Valley System*, p. 9.

CHAPTER FOURTEEN. MEXICAN GENESES

82 "From an evolutionary standpoint": Axelrod, "Evolution of the Madro-Tertiary Geoflora," p. 458.
83 "Late Pliocene and Quaternary elevation": Ibid., p. 503.
84 "On the contrary, the phytogeographic data": Rzedowski, "Algunas Consideraciones Acerca del Elemento Endemico en la Flora de Mexico," p. 61.
85 "The floras of the middle and latter part": Stebbins, *Variation and Evolution in Plants*, p. 521.
85 "Of the genera studied": Rzedowski, "Algunas Consideraciones Acerca del Elemento Endemico en la Flora de Mexico," p. 62.
86 "The abundance of the endemic element": Ibid., p. 6.

CHAPTER FIFTEEN. DESERT RELICTS

88 "Finally, there belongs": Barco, *Natural History of Baja California*, p. 176.
89 "They are made of nothing else": Ibid., p. 177.
90 "Stebbins moved there": Crawford and Smocovitis, *Scientific Papers of G. Ledyard Stebbins*, p. 22.
90 "I have discussed": Mead, "Life and Work of G. Ledyard Stebbins," p. 48.
90 "So, my book": Ibid., p. 139.
90 "one in the Siskiyou-Trinity": Stebbins and Major, "Endemism and Speciation in the California Flora," p. 10.
91 "In the first place": Ibid., p. 13.
92 "Stebbins and Major thought it 'highly unlikely'": Ibid., p. 14.
92 "During the early part of the Tertiary": Ibid., p. 15.

CHAPTER SIXTEEN. MADRO-TERTIARY ATTITUDES

95 "The distribution of forests": Axelrod, "Fossil Floras Suggest Stable, Not Drifting Continents," p. 3257.

96 "The recent discovery": Axelrod, "Ocean Floor Spreading in Relation to Ecosystematic Problems," p. 15.

96 "Evolution of angiosperms": Axelrod, "Drought, Diastrophism, and Quantum Evolution," p. 202.

97 "Stebbins (1952) has shown": Axelrod, "Edaphic Aridity as a Factor in Angiosperm Evolution," p. 311.

97 "The domelands of the southern Sierra": Ibid., p. 315.

98 "Although he was not always right": Lipps, "Daniel Isaac Axelrod (1910–1998)."

98 "He would often pull me aside": Anonymous, "Distinguished Botanist, Daniel Axelrod, Dies at 87."

98 "Except for a few species": A. W. Johnson, "Evolution of Desert Vegetation in North America," p. 135.

98 "Many of these taxa": Ibid.

99 "In general, one may conclude": Ibid., p. 129.

CHAPTER SEVENTEEN. A FRIENDLY LAND

101 "One of the noisier recent struggles": "Judgment Day: Intelligent Design on Trial," *Nova*, PBS, November 13, 2007.

101 "Our beautiful deserts": Jaeger, *California Deserts*, p. 189.

102 "It has become a habit": Ibid., p. 184.

102 "The chuckwalla is a thoroughgoing vegetarian": Ibid., p. 76.

103 "It is indeed possible": Ibid., p. 125.

103 "In masses of conglomerate": Ibid., p. 30.

103 "Desert plants commonly exhibit": Ibid., p. 126.

104 "With every passing year": Krutch, *Voice of the Desert*, p. 148.

104 "How does it happen that this striking creature": Krutch, *Forgotten Peninsula*, p. 212.

104 "One must imagine": Ibid., p. 214.

104 "Just how long have our deserts been deserts?": Ibid., p. 52.

105 "He quoted the same passage": Ibid., p. 53.

105 "Though very little plant fossil material": Ibid., p. 50.

105 "The cactus originated": Krutch, *Voice of the Desert*, p. 63.

106 "He seemed unaware": Krutch, *Baja California and the Geography of Hope*, p. 122.

106 "If it is really a prickly pear": Krutch, *Voice of the Desert*, p. 58.

106 "Contemplating ocotillo's giant relative": Krutch, *Forgotten Peninsula*, p. 91.

106 "something like what it now is": Krutch, *Voice of the Desert*, p.55.

107 "Most unfortunate it is": Jaeger, *California Deserts*, p. 189.

107 "Baja California is a wonderful example": Krutch, *Baja California and the Geography of Hope*, p. 10.

107   "such creatures as the scorpions": Krutch, *Desert Year*, p. 42.
107   "As for the animals": Ibid., p. 63.
107   "To those who do listen": Krutch, *Voice of the Desert*, p. 220.

CHAPTER EIGHTEEN. FURRY PALEONTOLOGISTS

110   "So we turned up a cañon": Manly, *Death Valley in '49*, p. 77.
111   "One historian thought": Chalfant, *Death Valley*, p. 78.
111   "Later observers were convinced": Manly, *Death Valley in '49*, p. 343 n. 20.
112   "As they sat and commiserated": Betancourt, Van Devender, and Martin, *Packrat Middens*, p. 3.
112   "The limited foraging range": Wells and Jorgensen, "Pleistocene Wood Rat Middens and Climatic Change in Mohave Desert," p. 1172.
112   "Seventeen ancient wood rat middens": Wells and Berger, "Late Pleistocene History of Coniferous Woodland in the Mohave Desert," p. 1646.
113   "Prevalence of woodland vegetation": Ibid., p. 1641.
113   "That regional climate": Axelrod, *Quaternary Extinctions of Large Mammals*, p. 13.
113   "a modest radiation in the late Miocene": Betancourt, Van Devender, and Martin, *Packrat Middens*, p. 15.
114   "long distance migration": A. W. Johnson, "Evolution of Desert Vegetation in Western North America," p. 298.
115   "The striking cytogeographic differentiation": Wells and Hunziker. "Origin of the Creosote Bush *(Larrea)* Deserts of Southwestern North America," p. 853.
115   "Outlying pockets of *Larrea*": Barbour, "Patterns of Genetic Similarity Between *Larrea tridentata* in North and South America," p. 66.
116   "The meager fossil record": Raven and Axelrod, "Angiosperm Biogeography and Past Continental Movements," p. 630.

CHAPTER NINETEEN. DAWN HORSES AND DINOSAURS

117   "The oldest continental Cenozoic": McKenna, "Continental Paleocene Fauna from California," p. 1.
118   "extensive prospecting efforts": Ibid., p. 3
118   "from the water well behind the cabin": McKenna, "Paleocene Mammals, Goler Formation, Mojave Desert, California," p. 512.
118   "The Mohave [now Goler] formation": Axelrod, "Eocene and Oligocene Formations of the Western Great Basin," p. 1935.

118   "Prospecting has been intermittently continued": McKenna, "Continental Paleocene Fauna from California," p. 4.

119   "of some reptile adapted to powerful digging": Ibid., p. 6.

120   "an abundance of petrified wood": Morris, "Baja California: Late Cretaceous Dinosaurs," p. 1539.

120   "The lithology of the El Gallo Formation": Ibid., p. 1540.

120   "could be compared favorably": Novacek, *Time Traveler,* p. 180.

121   "The only plant fossils cited": Novacek et al., "Wasatchian (Early Eocene) Mammals and Other Vertebrates from Baja California, Mexico," p. 9.

121   "When petrified wood from the Miocene": McKeown, Luz, and Jones, "Fossil Wood from the Miocene Comondu Formation," p. 7.

CHAPTER TWENTY. AXELROD ANTAGONISTES

122   "to present the evidence": G. Ledyard Stebbins, letter of May 10, 1983.

123   "originated from a group": Sherwin Carlquist, letter of June 1, 1983.

123   "I do not wish to get involved": Daniel Axelrod, letter of May 25, 1983.

123   "sometimes gruff, blunt demeanor": Anonymous, "Distinguished Botanist, Daniel Axelrod, Dies at 87."

123   "Axelrod's model": Sarmiento, "Arid Vegetation in Tropical America," p. 95.

124   "Desert adaptation": Blair, Hulse, and Mares, "Origins and Affinities of the Vertebrates of the North American Sonoran Desert and the Monte Desert of Northwestern Argentina," p. 1.

124   "Several observations suggest": Otte, "Species Richness Patterns of New World Desert Grasshoppers in Relation to Plant Diversity," p. 208.

125   "It seems then that selection": Blair, "Adaptations of Anurans to Equivalent Desert Scrub in North and South America," p. 202.

126   "After his retirement": Lipps, "Daniel I. Axelrod (1910–1998)."

126   "The tadpoles transform": Axelrod, "Evolution and Biogeography of Madrean-Tethyan Sclerophyll Vegetation," p. 314.

126   "sufficiently abundant in the late middle Eocene": Ibid., p. 313.

126   "Is zonal desert vegetation ancient": Axelrod, "Desert Vegetation," p. 1.

127   "Since the lowlands": Axelrod, "Age and Origin of Sonoran Desert Vegetation," p. 52.

128   "As an alternative, it seems more probable": Ibid., p. 55.

128   "The Monte has the greatest taxonomic diversity": Blair, "Adaptations of Anurans to Equivalent Desert Scrub in North and South America, p. 216.

128   "The greater taxonomic diversity": Ibid., p. 208.

128   "Paleobotanical evidence suggests": Ibid., p. 216.

CHAPTER TWENTY-ONE. THE MIDDAY SUN

130 "occupying habitats associated with exposures": Marlow, Brody, and Wake, "New Salamander, Genus *Batrachoseps,* from the Inyo Mountains of California," p. 16.

132 "almost defy the laws of physics": Greene, *Snakes,* p. 36.

132 "the farm fence lizards": Ballou, "Serpentlike Sea Saurians," p. 223.

133 "I spent a number of days": Cowles, *Desert Journal,* p. 91.

133 "You are old": Carroll, "Father William," in *Alice's Adventures in Wonderland,* p. 51.

134 "an occasional and very brief 113 degrees F": Cowles, *Desert Journal,* p. 84.

135 "But he had described showy orange": Keynes, *Charles Darwin's Zoology Notes and Specimen Lists from H.M.S. Beagle,* p. 295.

135 "The lizards are many": Van Dyke, *Desert,* p. 170.

135 "having traced with surprise": Ibid., p. 173.

138 "Besides the advances and retreats": Krutch, *Desert Year,* p. 45.

138 "Ears comprise the reception system": Pianka and Vitt, *Lizards,* p. 93.

138 "showed an enormous capacity": Rowntree, *Hardy Californians,* p. 15.

CHAPTER TWENTY-TWO. LACERTILIAN AMBIGUITIES

139 "The performance went on": Krutch, *Desert Year,* p. 45.

142 "More recently, a landfill": Kelley et al., "Preliminary Report of a Paleontological Investigation of the Lower and Middle Members, Sespe Formation, Simi Valley Landfill, Ventura County, California."

142 "And renewed digging": Lofgren et al., "Paleocene Primates from the Goler Formation of the Mojave Desert in California," p. 11.

142 "Every quarter mile": Chase, *California Desert Trails,* p. 276.

143 "During the Pliocene": Jefferson and Lindsay, *Fossil Treasures of the Anza-Borrego Desert,* p. 84.

143 "The Anza Borrego lizard assemblage": Norell, "Late Cenozoic Lizards of the Anza Borrego Desert, California," p. 28.

144 "It was shown that the composition": Holman, *Pleistocene Amphibians and Reptiles in North America,* p. 198.

145 "The analysis of these specimens": Norell, "Late Cenozoic Lizards of the Anza Borrego Desert, California," p. 28.

CHAPTER TWENTY-THREE. XEROTHERMIC INVASIONS

147 "perhaps the most remarkable of all desert birds": Krutch, *Voice of the Desert*, pp. 33, 36–37.
148 "Formerly the range of the road runner": Jaeger, *Desert Wildlife*, p. 189.
149 "A unique type of vegetation": Stebbins and Taylor, *Survey of the Natural History of the South Pacific Border Region*, p. 153.
149 "A significant, readily recognizable warming trend": Axelrod, "Pleistocene Soboba Flora of Southern California," p. 42.
150 "Coville (1893) noted the occurrence": Ibid, p. 46.

CHAPTER TWENTY-FOUR. SAND SWIMMERS

153 "In the first place, the body": Schmidt and Inger, *Living Reptiles of the World*, p. 116.
153 "The specializations that allow *Uma*": Pough, Morafka, and Hillman, "Ecology and Burrowing Behavior of the Chihuahuan Fringe-Footed Lizard, *Uma exsul*," p. 85.
153 "The Chihuahuan Desert cradles": Morafka, "Biogeographical Analysis of the Chihuahuan Desert through Its Herpetofauna," pp. 174, 175.
154 "The eastern or Chihuahuan desert": Morafka and Reyes, "Biogeography of Chihuahuan Desert Herpetofauna," p. 85.
154 "Vegetation in the [Mapimian] region": Morafka, "Biogeographical Analysis of the Chihuahuan Desert through Its Herpetofauna," p. 174.
155 "The underlying cause": Morafka and Reyes, "Biogeography of Chihuahuan Desert Herpetofauna," p. 85.

CHAPTER TWENTY-FIVE. AXELROD ASCENDANT

156 "A number of genera": Raven and Axelrod, *Origin and Relationships of the California Flora*, p. 45.
156 "Madro-Tertiary vegetation": Ibid., pp. 21, 46.
157 "the first person who matched": Barbour, "Dan Axelrod: Paleoecologist for the Ages," p. 29.
157 "The present California desert flora": Thorne, "Historical Sketch of the Vegetation of the Mojave and Colorado Deserts of the American Southwest," pp. 647, 648.
158 "If our present California deserts": Ibid., p. 645.
158 "Those of us without much geological training": Ibid., p. 642.
158 "The erosion that stripped": Norris and Webb, *Geology of California*, p. 86.

159 "After the batholith": McPhee, *Assembling California*, p. 31.
159 "This brief summary": Axelrod, "Paleobotanical History of the Western Deserts," p. 129.

CHAPTER TWENTY-SIX. AN EVOLUTIONARY MUSEUM

160 "nearly all existed": Stebbins, "Aridity as a Stimulus to Plant Evolution," p. 33.
161 "two of the principal groups": Stebbins, *Flowering Plants*, p. 251.
162 "radiating complexes": Ibid., p. 164.
162 "Some scientists have tried": Schaffer, "California's Geological History and Changing Landscapes," p. 49.
163 "Daniel Axelrod in 1957": Schaffer, *Geomorphic Evolution of the Yosemite Valley and Sierra Nevada Landscapes*, p. 303.
163 "Desert landscapes are the most ancient": Sokolov, Haffter, and Ortega, *Vertebrate Ecology in Arid Zones of Mexico and Asia*, p. 12.
163 "little hope that such fossils have been preserved": Rzedowski, "Diversidad y Origines de la Flora Fanerogamica de Mexico," p. 19.
164 "This phenomenon is particularly spectacular": Ibid., p. 14.

CHAPTER TWENTY-SEVEN. THE RIDDLE OF THE PALMS

165 "Under the palms": King, *Mountaineering in the Sierra Nevada*, p. 37.
167 "partly hidden among dunes": Chase, *California Desert Trails*, p. 93.
168 "Here, on November 29": Jaeger, *California Deserts*, p. 180.
168 "These are the so-called oases": J. C. Van Dyke, *Desert*, p. 35.
168 "Some of the groups occur": Chase, *California Desert Trails*, p. 16.
169 "definitely known to be a native": Jaeger, *Desert Wild Flowers*, p. 6.
169 "closely similar fossils": Axelrod, "Evolution of Desert Vegetation in Western North America," p. 271.
169 "It is less obvious": Cornett, "Factors Determining the Occurrence of the Desert Fan Palm, *Washingtonia filifera*," p. 37.
169 "three a-priori predictions": McClenaghan and Beauchamp, "Low Genic Differentiation among Isolated Populations of California Fan Palm (*Washingtonia filifera*)," p. 316.
170 "A more plausible scenario": Ibid., p. 322.
170 "There does not, in fact, appear to be any fossil evidence": Cornett, "Desert Fan Palm," p. 56.
171 "These four lines of evidence": Ibid., p. 57.
171 "the desert fan palm": Cornett, "Population Dynamics of the Palm, *Washingtonia filifera*, and Global Warming," p. 47.

CHAPTER TWENTY-EIGHT. BUSHES AND CAMELS

173   "It is likely that contemporary": Janzen, "Chihuahuan Desert Nopaleras," p. 625.

174   "There are many books": Ibid., p. 597.

174   "There were four genera": Ibid., p. 611.

174   "In the absence of contemporary": Ibid., p. 616.

175   *Acacia farnesiana* clearly": Daniel Janzen, e-mail of February 16, 2008.

176   "Then why the bright colors": Janzen, "Chihuahuan Desert Nopaleras," p. 613.

176   "He cited the presence": Ibid., p. 611.

176   "Jumping cholla may well be the nastiest": Ibid., p. 625.

176   "During recent fieldwork": Janzen, "Depression of Reptile Biomass by Large Herbivores," p. 381.

177   "The ease with which leaf eating": Ibid., p. 389.

177   "a study in the Kalahari": Ibid., p. 390.

177   "Cacti are widely believed": Janzen, "Chihuahuan Desert Nopaleras," p. 599.

178   "The plant-megafaunal interactions": Ibid., p. 623.

178   "Ocotillo and cacti": Daniel Janzen, e-mail of February 15, 2008.

CHAPTER TWENTY-NINE. AXELROD ASKEW

181   "How doth the little crocodile": Carroll, *Alice's Adventures in Wonderland*, p. 26.

182   "During an era": Barbour, "Dan Axelrod: Paleoecologist for the Ages," p. 29.

183   "The Sierra may have maintained": Wernicke et al., "Origin of the High Mountains in the Continents," p. 192.

183   "A year earlier": Dokka and Ross, "Collapse of Southwestern North America and the Evolution of Early Miocene Detachment Faults, Metamorphic Core Complexes, the Sierra Nevada Orocline, and the San Andreas Fault System."

183   "Terrestrial plants are generally regarded": Wolfe et al., "Paleobotanical Evidence for High Altitudes in Nevada During the Miocene," p. 1672.

184   "Paleobotanical evidence supports": Ibid., p. 1674.

184   "Howard Schorn, a very meticulous paleobotanist": Schaffer, *Geomorphic Evolution of the Yosemite Valley and Sierra Nevada Landscapes: Solving the Riddles of the Rocks*, p. 303.

184   "We conclude that the Sierra ": House, Wernicke, and Farley, ""Dating Topography of the Sierra Nevada, California, Using Apatite (U-Th)/He Ages," p. 68.

185 "Because he believed": Schaffer, "Letters to the Editor," p. 31.

185 "Volcanic ashes currently exposed": Poage and Chamberlain, "Stable Isotopic Evidence for a Pre-Middle-Miocene Rain Shadow in the Western Basin and Range," p. 1.

186 "The paper compared 30-million-year-old": Retallack, Wynn, and Fremd, "Glacial-Interglacial-Scale Paleoclimatic Change without Large Ice Sheets in the Oligocene of Central Oregon," p. 298.

186 "The data, compared with modern isotopic": Mulch, Graham, and Chamberlain, "Hydrogen Isotopes in Eocene River Gravels and Paleoelevation of the Sierra Nevada," p. 87.

186 "Another *Science* article": Schuster et al., "Age of the Sahara Desert," p. 821.

CHAPTER THIRTY. PARADIGMS POSTPONED

187 "Recent geologic data": Stock, Anderson, and Finkel, "Pace of Landscape Evolution of the Sierra Nevada, California, Revealed by Cosmogenic Dating of Cave Sediments," p. 193.

188 "in a pattern that steepened": Ibid., p. 196.

188 "Low temperature geochronology studies": Ibid., p. 193.

189 "Prior to the rise": Pavlik, *California Deserts*, p. 130.

190 "This may represent the first documented fossil": Jefferson and Lindsay, *Fossil Treasures of the Anza-Borrego Desert*, p. 84.

192 "In this region the ground": Xenophon, *Anabasis*, p. 91.

193 "exactly what drove this": Pianka and Vitt, *Lizards*, p. 60.

CHAPTER THIRTY-ONE. THE FALCON AND THE SHRIKES

198 "He rode slowly home": Steinbeck, *To a God Unknown*, in *Novels and Stories*, pp. 325, 344.

198 "In the early spring": Chalfant, *Death Valley*, p. 6.

199 "There is something uneasy": Didion, *We Tell Ourselves Stories in Order to Live*, pp. 162, 224.

199 "Our regional deserts": Axelrod, "Paleobotanical History of the Western Deserts," p. 129.

200 "My sensibility was transfixed": Banham, *Scenes In America Deserta*, p. 11.

200 "The cliché 'man-made desert'": Ibid., p. 204.

200 "Ultimately deserts are man-made": Ibid., p. 205.

201 "to hear with any inner understanding" : Ibid., p. 164.

202 "In calling up images": Darwin, *Voyage of the Beagle*, p. 484.

EPILOGUE. THE SPHINX'S LAIR

205 "The small Ricardo flora": Axelrod, "Paleobotanical History of the Western Deserts," p. 122.

207 "Things had a way of happening": Austin, *Earth Horizon*, p. 258.

207 "The palpable sense of mystery": Austin, *Land of Little Rain*, in *Stories from the Country of Lost Borders*, p. 16.

208 "Environmentalists say desert is so hostile": Darlington, *Mojave*, p. 163.

208 "Antienvironmentalists say desert is so hostile": Fife and Dickey, "Fragile Soils of the California Desert."

208 "Joseph Chase was already complaining": Chase, *California Desert Trails*, p. 37.

208 "ORVs scraped twelve thousand tons": Darlington, *Mojave*, p. 254.

209 "What monstrous folly": J.C. Van Dyke, *Desert*, p. xvii.

209 "Current attempts and plans": Axelrod, "Desert Vegetation," p. 57.

210 "They say the Lion": Jamshyd was a ruler of the mythical first Persian dynasty, the Pishdadians. The ruins of Persepolis, the fourth century B.C. Achaemenid Dynasty's capital, are known as Jamshyd's Throne. Bahram V, famed for hunting and amorous exploits, was a ruler of Persia's A.D. fifth century Sasanian Empire. He was nicknamed Bahram Gor— Bahram the Wild Ass.

Selected Bibliography

Abbey, Edward. *Cactus Country*. New York: Time-Life Books, 1973.
————. *The Journey Home*. New York: Dutton, 1977.
Anonymous. "Distinguished Botanist, Daniel Axelrod, Dies at 87." *UC Davis News and Information*, June 4, 1998.
Austin, Mary. *Earth Horizon*. Boston: Houghton Mifflin, 1932.
————. *Stories from the Country of Lost Borders*. New Brunswick, New Jersey: Rutgers University Press, 1987.
Axelrod, Daniel I. "A Pliocene Flora from the Eden Beds." *American Museum Novitates*, no. 729 (June 6, 1934).
————. "A Pliocene Flora from the Mount Eden Beds, Southern California." *Carnegie Institution of Washington Publication* 476 (1937): 125–183.
————. "A Miocene Flora from the Western Border of the Mohave Desert." *Carnegie Institution of Washington Contributions to Paleontology* 156 (1939).
————. "Eocene and Oligocene Formations of the Western Great Basin." *Bulletin of the Geological Society of America* 60 (1949): 1935–1936.

——. "Evolution of Desert Vegetation in Western North America." *Carnegie Institution of Washington Contributions to Paleontology* 590 (1950).

——. "Evolution of the Madro-Tertiary Geoflora." *The Botanical Review* 24 (July 1958): 433–509.

——. "Fossil Floras Suggest Stable, Not Drifting Continents." *Journal of Geophysical Research* 68 (May 15, 1963): 3257–3263.

——. "The Pleistocene Soboba Flora of Southern California." *University of California Publications in Geological Science* 60 (1966).

——. "Drought, Diastrophism, and Quantum Evolution." *Evolution* 21 (June 1967): 201–209.

——. "Quaternary Extinctions of Large Mammals." *University of California Publications in Geological Science* 74 (1967).

——. "Mesozoic Paleogeography and Early Angiosperm History." *The Botanical Review* 36 (July–September 1970): 277–319.

——. "Edaphic Aridity as a Factor in Angiosperm Evolution." *The American Naturalist* 106 (May–June 1972): 311–320.

——. "Ocean Floor Spreading in Relation to Ecosystematic Problems." In *A Symposium in Ecosystematics,* edited by Robert T. Allen and Frances C. James. *University of Arkansas Occasional Paper* 4 (1972): 15–76.

——. "Evolution and Biogeography of Madrean-Tethyan Sclerophyll Vegetation." *Annals of the Missouri Botanical Garden* 62 (1975): 280–334.

——. "Age and Origin of Sonoran Desert Vegetation." *Occasional Papers of the California Academy of Sciences* 132 (June 12, 1979).

——. "Desert Vegetation: Its Age and Origin." In *Arid Land Plant Resources: Proceedings of the International Arid Lands Conference on Plant Resources, Texas Tech University,* edited by J. R. Goodin and David K. Northington. Lubbock, Texas: International Center for Arid and Semi-Arid Land Studies, July 1979.

——. "Paleobotanical History of the Western Deserts." In *Origins and Evolution of Deserts: Symposium of the Committee on Desert and Arid Zones Research of the Southwestern and Rocky Mountain Division of the American Association of the Advancement of Science,* edited by Stephen G. Wells and Daniel R. Haragan. Albuquerque: University of New Mexico Press, 1983.

——. "Response of Daniel I. Axelrod for the Award of the Paleontological Society Medal." *Journal of Paleontology* 65 (1991): 522–523.

Ballou, William Hosea. "The Serpentlike Sea Saurians." *Appleton's Popular Science Monthly* 53 (1898): 209–225.

Banham, Peter Reyner. *Scenes in America Deserta.* Salt Lake City: Peregrine Smith Books, 1982.

Barbour, Michael G. "Patterns of Genetic Similarity Between *Larrea tridentata* in North and South America." *American Midland Naturalist* 81 (1969): 54–58.

——. "Ecological Fragmentation in the Fifties." In *Uncommon Ground: Toward*

Reinventing Nature, edited by William Cronon. New York: W. W. Norton, 1995.

———. "Dan Axelrod: Paleoecologist for the Ages." *Fremontia* 27 (January 1999): 29–30.

Barco, Miguel del. *Historia Natural y Cronica de la Antigua California.* Edited by Miguel Leon Portilla. Mexico City: Universidad Nacional Autonoma de Mexico, 1973.

———. *The Natural History of Baja California.* Translated by Froylan Tiscareno. Los Angeles: Dawson's Book Shop, 1980.

Beidleman, Richard G. "Rowboat Botanizing with Willis Linn Jepson on the Colorado River, 1912." *Fremontia* 28 (April–October 2000): 3–12.

———. "Willis Lynn Jepson: 'The Botany Man.'" *Madrono* 47 (2000): 273–286.

Betancourt, Julio L., Thomas R. Van Devender, and Paul S. Martin. *Packrat Middens: The Last 40,000 Years of Biotic Change.* Tucson: University of Arizona Press, 1990.

Blair, W. Frank. "Adaptations of Anurans to Equivalent Desert Scrub in North and South America." In *Evolution of Desert Biota,* edited by David W. Goodall. Austin: University of Texas Press, 1976.

Blair, W. Frank., Arthur C. Hulse, and Michael A. Mares. "Origins and Affinities of the Vertebrates of the North American Sonoran Desert and the Monte Desert of Northwestern Argentina." *Journal of Biogeography* 3 (1976): 1–18.

Bowers, Janice Emily. *A Sense of Place: The Life and Work of Forrest Shreve.* Tucson: University of Arizona Press, 1988.

Brewer, William H. *Up and Down California.* Berkeley: University of California Press, 1966.

Cain, S. A. *Foundations of Plant Geography.* New York: Harper and Brothers, 1944.

Campbell, Joseph. *The Masks of God: Oriental Mythology.* New York: Viking Press, 1962.

Carroll, Lewis. *Alice's Adventures in Wonderland.* New York: Airmont Publishing Company, 1965.

Chalfant, W. A., *Death Valley: The Facts.* Stanford, California: Stanford University Press, 1936.

Chase, Joseph Smeaton. *California Desert Trails.* Palo Alto, California: Tioga Publishing Company, 1987.

Clements, Frederic E. "The Origin of the Desert Climax and Climate." In *Essays in Geobotany in Honor of William Albert Setchell,* edited by T. H. Goodspeed. Berkeley: University of California Press, 1936.

Cope, Edward D. *On the Method of the Creation of Organic Types.* Pamphlet. Philadelphia: M'Calla and Stavely, 1871.

———. "The Reptiles of the American Eocene." *The American Naturalist* 16 (1882): 979–993.

————. *The Crocodilians, Lizards, and Snakes of North America.* In *Report of the U.S. National Museum,* 153–1270. Washington, D.C., 1898.

Cornett, James W. "The Desert Fan Palm: Not a Relict." Abstracts of the Proceedings, 1989 Mojave Desert Quaternary Research Symposium. *San Bernardino County Museum Association Quarterly* 36 (Summer 1989): 56–58.

————. *Desert Palm Oasis.* Palm Springs, California: Palm Springs Desert Museum, 1989.

————. "Population Dynamics of the Palm, *Washingtonia filifera,* and Global Warming." Abstracts of the Proceedings from the Fifth Annual Mojave Desert Quaternary Research Symposium, May 17–18, 1991. *San Bernardino County Museum Association Quarterly* 38 (Summer 1991): 46–47.

————. "Factors Determining the Occurrence of the Desert Fan Palm, *Washingtonia filifera.*" *San Bernardino Country Museum Association Special Publication* 93 (1993): 37–38.

Coville, F. V. *Botany of the Death Valley Expedition.* Contributions to the U.S. National Herbarium 14. Washington, D.C.: Government Printing Office, 1893.

Cowles, Raymond B. *Desert Journal: A Naturalist Reflects on Arid California.* Berkeley: University of California Press, 1977.

Crawford, Daniel J., and Betty Vasiliki Smocovitis, editors. *The Scientific Papers of G. Ledyard Stebbins (1929–2000).* Liechtenstein: A. R. G. Gatner Verlag, 2004.

Darlington, David. *The Mojave: A Portrait of the Definitive American Desert.* New York: Henry Holt, 1996.

Darwin, Charles. *The Origin of Species and The Descent of Man.* New York: Random House, Modern Library, no date.

————. *The Voyage of the Beagle.* London: J. M. Dutton, 1959.

————. *The Correspondence of Charles Darwin.* Vol. 6, *1856–1857.* Edited by Frederick Burkhardt and Sydney Smith. Cambridge: Cambridge University Press, 1990.

Deevey, E. S., Jr. "Biogeography of the Pleistocene." *Geological Society of America Bulletin* 60 (1949): 1315–1416.

Didion, Joan. *We Tell Ourselves Stories in Order to Live: Collected Nonfiction.* New York: Alfred A. Knopf, 2006.

Dokka, Roy, and Timothy M. Ross. "Collapse of Southwestern North America and the Evolution of Early Miocene Detachment Faults, Metamorphic Core Complexes, the Sierra Nevada Orocline, and the San Andreas Fault System." *Geology* 23 (1995): 1075–1078.

Fife, Donald L., and Robert H. Dickey. "Fragile Soils of the California Desert: Fact or Fiction?" *San Bernardino County Museum Association Quarterly* 37 (Summer 1990): 26.

Flynn, John J., and Michael J. Novacek. "Early Eocene Vertebrates from Baja California: Evidence for International Age Correlations." *Science* 224: 151–153.

Font, Pedro. *The Anza Expedition of 1775–1776*. Edited by Frederick Taggart. *Publications of the Academy of Pacific Coast History* 3 (March 1913).

Fremont, John C. *Report of the Exploring Expedition to the Rocky Mountains in the Year 1842 and to Oregon and North California in the Years 1843–44*. Washington, D.C.: Gales and Seaton, 1845.

Fremont, John C., and W. H. Emory. *Notes of Travel in California*. New York: D. Appleton and Company, 1849.

Gilmore, Charles W. *Fossil Lizards of North America*. Memoirs of the National Academy of Sciences 22. Washington, D.C.: Government Printing House, 1928.

Goodman, Susan, and Carl Dawson. *Mary Austin and the American West*. Berkeley: University of California Press, 2008.

Gould, Stephen Jay. *Time's Arrow, Time's Cycle: Myth and Metaphor in the Discovery of Geological Time*. Cambridge, Massachusetts: Harvard University Press, 1987.

Gray, Asa. "Characteristics of the North American Flora." *American Journal of Science and Arts* 3, no. 28 (1884): 323–340.

Greene, Harry W. *Snakes: The Evolution of Mystery in Nature*. Berkeley: University of California Press, 1997.

Heizer, Robert. *Handbook of North American Indians*. Vol. 8. Washington, D.C.: Smithsonian Institution, 1983.

Henslow, Rev. George. *The Origin of Plant Structures*. London: Kegan, Paul, Trench, Trubner and Company, 1895.

Holman, J. Alan. *Pleistocene Amphibians and Reptiles in North America*. Oxford: Oxford University Press, 1995.

House, Martha A., Brian P. Wernicke, and Kenneth A. Farley. "Dating Topography of the Sierra Nevada, California, Using Apatite (U-Th)/He Ages." *Nature* 396 (1998): 66–69.

Howard, Richard A. "Ivan Murray Johnston, 1898–1960." *Journal of the Arnold Arboretum* 42 (January 1961): 1–9.

Jaeger, Edmund C. *Desert Wild Flowers*. Stanford, California: Stanford University Press, 1940.

———. *Desert Wildlife*. Stanford, California: Stanford University Press, 1961.

———. *The California Deserts*. 4th edition. Stanford, California: Stanford University Press, 1965.

———. *A Naturalist's Death Valley*. Revised edition. San Bernardino, California: Death Valley 49'ers, 1968.

Janzen, Daniel. "The Depression of Reptile Biomass by Large Herbivores." *The American Naturalist* 110 (1976): 371–400.

———. "Chihuahuan Desert Nopaleras: Defaunated Big Mammal Vegetation." *Annual Review of Ecological Systems* 17 (1986): 595–636.

Jefferson, George T., and Lowell Lindsay. *Fossil Treasures of the Anza-Borrego Desert*. San Diego: Sunbelt Publications, 2006.

Jepson, Willis Linn. *A Manual of the Flowering Plants of California*. Berkeley: University of California Press, 1951.

Johnson, A. W. "The Evolution of Desert Vegetation in North America." In *Desert Biology*, vol. 1, edited by G. W. Brown. New York: Academic Press, 1968.

Johnson, David H., Monroe D. Bryant, and Alden Miller. "Vertebrate Animals of the Providence Mountains Area of California." *University of California Publications in Zoology* 48 (1948): 221–376.

Johnston, Ivan M. "The Floristic Significance of Shrubs Common to the North and South American Deserts." *Journal of the Arnold Arboretum* 21 (July 1940): 356–363.

Jones, Lawrence L. C., and Robert E. Lovitch, editors. *Lizards of the American Southwest*. Tucson, Arizona: Rio Nuevo Publishers, 2009.

Kelley, Thomas S., E. Bruce Lander, David P. Whistler, Mark A. Roeder, and Robert E. Reynolds. "Preliminary Report of a Paleontological Investigation of the Lower and Middle Members, Sespe Formation, Simi Valley Landfill, Ventura County, California." *Paleobios* 13 (July 26, 1991): 1–13.

Keynes, Richard D., editor. *Charles Darwin's Zoology Notes and Specimen Lists from H.M.S. Beagle*. Cambridge: Cambridge University Press, 2000.

King, Clarence. *Mountaineering in the Sierra Nevada*. New York: W. W. Norton and Company, 1935.

Kowalewski, Michael, editor. *Reading the West: New Essays on the Literature of the American West*. Cambridge: Cambridge University Press, 1996.

Krutch, Joseph Wood. *The Voice of the Desert: A Naturalist's Interpretation*. New York: William Sloan Associates, 1954.

———. *The Forgotten Peninsula: A Naturalist in Baja California*. New York: William Sloan Associates, 1961.

———. *The Desert Year*. New York: Viking Press, 1963.

———. *Baja California and the Geography of Hope*. San Francisco: Sierra Club Books, 1967.

Laird, Carobeth. *The Chemehuevis*. Banning, California: Malki Museum Press, 1976.

Lancaster, Lesley T. "Adaptive Social and Maternal Induction of Anitpredator Dorsal Patterns in a Lizard with Alternative Social Strategies." *Ecology Letters* 10 (June 10, 2007): 798–808.

Lee, W. Storrs. *The Great California Deserts*. New York: G. P. Putnam's Sons, 1963.

Lema, Sean C. "The Phenotypic Plasticity of Death Valley's Pupfish." *American Scientist* 96 (January–February 2008): 28–36.

Lipps, Jere H. "Daniel Isaac Axelrod (1910–1998)." *Botanical Electrical News* 202 (September 12, 1998).

Lofgren, Donald, James G. Honey, Malcolm C. McKenna, Robert L. Zondervan, and Erin E. Smith. "Paleocene Primates from the Goler Formation of the Mojave Desert in California." In *Geology and Vertebrate Paleontology of Western and Southern North America*, edited by Xiaoming Wang and Lawrence G. Barnes. Science Series 41. Los Angeles: Natural History Museum of Los Angeles Country, May 28, 2008.

Lurie, Edward. *Louis Agassiz: A Life in Science*. Chicago: University of Chicago Press, 1960.

MacDougal, Daniel Trembly. *Botanical Features of North American Deserts*. Washington, D.C.: Carnegie Institution of Washington, 1908.

————. "Influence of Aridity upon the Evolutionary Development of Plants." *The Plant World* 12 (October 1909): 217–231.

————. "The Origin of Desert Floras." In *Distribution and Movements of Desert Plants*, by V. M. Spalding. Publication no. 113. Washington, D.C.: Carnegie Institution of Washington, 1909.

Manly, William Lewis. *Death Valley in '49*. Berkeley, California: Heyday Books, 2001.

Marlow, Ronald William, John M. Brody, and David Wake. "A New Salamander, Genus *Batrachoseps*, from the Inyo Mountains of California, with a Discussion of Relationships in the Genus." *Contributions to Science of the Natural History Museum of Los Angeles* 308 (March 16, 1979): 1–17.

Marsh, Othniel Charles. "Preliminary Description of New Tertiary Lizards." *American Journal of Science* 4 (October 1872): 298–309.

McClenaghan, Leroy R. J., and Arthur C. Beauchamp. "Low Genic Differentiation among Isolated Populations of California Fan Palm *(Washingtonia filifera)*." *Evolution* 40 (1986): 315–322.

McKenna, Malcolm C. "Paleocene Mammals, Goler Formation, Mojave Desert, California." *Bulletin of the American Association of Petroleum Geologists* 39 (April 1955): 512–515.

————. "A Continental Paleocene Fauna from California." *American Museum Novitates* no. 2024 (1960): 1–20.

McKeown, Leon de la Luz, and Jones. "Fossil Wood from the Miocene Comondu Formation of Baja California Sur." *Paleobios* 13 (April 12, 1991): 7.

McPhee, John. *Assembling California*. New York: Farrar, Straus and Giroux, 1993.

Mead, Mary. "Life and Work of G. Ledyard Stebbins: An Oral History." Special Collections Department, Shields Library, University of California, Davis, 1993.

Merriam, C. Hart. "Is Mutation a Factor in the Evolution of Higher Vertebrates?" *Science* 23 (February 16, 1906): 241–257.

Merriam, C. Hart, A. K. Fisher, Leonhard Stejneger, Charles H. Gilbert, C. V. Riley, R. E. C. Stearns, and T. S. Palmer. *The Death Valley Expedition*. North American Fauna no. 7. Washington D.C.: Government Printing Office, 1893.

Merriam, John C. "Extinct Faunas of the Mohave Desert, Their Significance in a Study of the Origin and Evolution of Life in America." *Popular Science Monthly* (March 1915): 245–264.

———. "Tertiary Mammalian Faunas of the Mohave Desert." *University of California Publications in Geology* 11 (August 1919).

Miller, James S., Mary Sue Taylor, and Erin Rempala. *Ivan M. Johnston's Studies in the Boraginaceae*. St. Louis: Missouri Botanical Garden Press, 2005.

Morafka, David J. "A Biogeographical Analysis of the Chihuahuan Desert through Its Herpetofauna." *Biogeographica* 9 (1977).

Morafka, David J., and Luz M. Reyes, "The Biogeography of Chihuahuan Desert Herpetofauna: Old Myths and New Realities." *Southwestern Herpetologists Society Special Publication*, no. 5 (October 1994): 79–87.

Morgan, Dale L. *Jedediah Smith and the Opening of the West*. Lincoln: University of Nebraska Press, 1964.

Morris, William J. "Fossil Mammals from Baja California: New Evidence of Early Tertiary Migrations." *Science* 153 (1966): 1376–1378.

———. "Baja California: Late Cretaceous Dinosaurs." *Science* 155 (1967): 1539–1541.

Muir, John. *The Mountains of California*. Charleston, South Carolina: BiblioBazaar, 2006.

Mulch, Andreas, Stephen A. Graham, and C. Page Chamberlain. "Hydrogen Isotopes in Eocene River Gravels and Paleoelevation of the Sierra Nevada." *Science* 313 (2006): 87–89.

Munz, Philip A., and David D. Keck. *A California Flora and Supplement*. Berkeley: University of California Press, 1968.

Norell, Mark A. "Late Cenozoic Lizards of the Anza Borrego Desert, California." *Contributions in Science of the Natural History Museum of Los Angeles County* 414 (1989).

Norris, Robert M., and Robert W. Webb. *Geology of California*. New York: John Wiley and Sons, 1990.

Novacek, Michael J. *Time Traveler*. New York: Farrar, Strauss and Giroux, 2002.

Novacek, Michael J., Ismael Ferrusquia-Villafranca, John J. Flynn, Andre R. Wyss, and Mark Norrell. "Wasatchian (Early Eocene) Mammals and Other Vertebrates from Baja California, Mexico: The Lomas Las Tetas de Cabra Formation." *Bulletin of the American Museum of Natural History* 208 (1991): 1–88.

Osborn, Henry Fairfield. *Cope: Master Naturalist*. Princeton, New Jersey: Princeton University Press, 1931.

Otte, Daniel. "Species Richness Patterns of New World Desert Grasshoppers in Relation to Plant Diversity." *Journal of Biogeography* 3 (1976): 197–209.

Patencio, Francisco. *Stories and Legends of the Palm Springs Indians.* Palm Springs, California: Palm Springs Desert Museum, 1943.

Pavlik, Bruce. *The California Deserts: An Ecological Rediscovery.* Berkeley: University of California Press, 2008.

Pianka, Eric R. *Ecology and Natural History of Desert Lizards.* Princeton, New Jersey: Princeton University Press, 1986.

Pianka, Eric R., and Laurie J. Vitt. *Lizards: Windows to the Evolution of Diversity.* Berkeley: University of California Press, 2003.

Poage, M. A., and C. P. Chamberlain. "Stable Isotopic Evidence for a Pre-Middle-Miocene Rain Shadow in the Western Basin and Range: Implications for the Paleotopography of the Sierra Nevada." *Tectonics* 21 (August 2002): 1–10.

Pough, F. Harvey, David J. Morafka, and Peter E. Hillman. "The Ecology and Burrowing Behavior of the Chihuahuan Fringe-Footed Lizard, *Uma exsul.*" *Copeia* (1978): 81–86.

Raven, Peter, and Daniel I. Axelrod, "Angiosperm Biogeography and Past Continental Movements." *Annals of the Missouri Botanical Garden* 61 (1974): 539–637.

———. *Origin and Relationships of the California Flora.* Berkeley: University of California Press, 1978.

Retallack, Gregory J., Jonathan G. Wynn, and Theodore J. Fremd. "Glacial-Interglacial-Scale Paleoclimatic Change without Large Ice Sheets in the Oligocene of Central Oregon." *Geology* 32 (April 2004): 297–300.

Rowntree, Lester. *Hardy Californians.* Berkeley: University of California Press, 2006.

Rzedowski, Jerzy. "Algunas Consideraciones Acerca del Elemento Endemico en la Flora de Mexico." *Boletin de la Sociedad Botanica de Mexico* 27 (November 1962): 47–63.

———. "Diversidad y Origines de la Flora Fanerogamica de Mexico." *Acta Botanica Mexicana* 14 (1991): 3–21.

———. "El Endemismo en la Flora Fanerogamica Mexicana." *Acta Botanica Mexicana* 15 (1991): 47–64.

Sarmiento, Guillermo. "Arid Vegetation in Tropical America." In *Evolution of Desert Biota,* edited by David W. Goodall. Austin: University of Texas Press, 1976.

Schaffer, Jeffrey P. "California's Geological History and Changing Landscapes." In *The Jepson Manual: Higher Plants of California,* edited by James C. Hickman. Berkeley: University of California Press, 1993.

———. *The Geomorphic Evolution of the Yosemite Valley and Sierra Nevada Landscapes: Solving the Riddles of the Rocks.* Berkeley: Wilderness Press, 1997.

———. "Letters to the Editor." *Fremontia* 27 (July 1999): 30–31.

Schmidt, Karl, and Robert F. Inger. *Living Reptiles of the World.* Garden City, New York: Doubleday and Company, 1957.

Schuster, Mathieu, Phillipe Duringer, Jean-Francois Ghienne, Patrick Vignaud, Hassan Taisso Mackaye, Androssa Likius, and Michel Brunet. "The Age of the Sahara Desert." *Science* 311 (2006): 821.

Shreve, Forrest. *The Cactus and Its Home.* Baltimore, Maryland: Williams and Wilkins, 1931.

Shreve, Forrest, and Ira L. Wiggins. *Vegetation of the Sonoran Desert.* Vol. 1. Publication no. 51. Washington D.C.: Carnegie Institution of Washington, 1951.

Sinervo, Barry. "Runaway Social Games, Genetic Cycles Driven by Alternative Male and Female Strategies, and the Origin of Morphs." *Genetica* 112 (2001): 417–443.

Sinervo, Barry, and Jean Colbert. "Morphs, Dispersal Behavior, Genetic Similarity, and Evolution of Cooperation." *Science* 300 (June 2003): 1949–1951.

Sinervo, Barry, Erik Svensson, and Tosha Commendant. "Density Cycles and Offspring Quantity and Quality." *Nature* 406 (2000): 985–988.

Sokolov, Vladimir, Gonzalo Haffter, and Alfredo Ortega. *Vertebrate Ecology in Arid Zones of Mexico and Asia.* Veracruz, Mexico: Instituto de Ecologia A.C., 1992.

Soltz, David L., and Robert J. Naiman, *The Natural History of Native Fishes in the Death Valley System.* Science Series 30. Los Angeles: Natural History Museum of Los Angeles County, 1979.

Spencer, William. "The Desert Fan Palm—Evidence Supports Relict Status." Internet article, 1995–2010. http://xeri.com/Moapa/relict.htm.

Stebbins, G. Ledyard. "Evidence of Rates of Evolution from the Distribution of Existing and Fossil Plant Species." *Ecological Monographs* 17 (April 1947): 149–158.

———. *Variation and Evolution in Plants.* New York: Columbia University Press, 1950.

———. "Aridity as a Stimulus to Plant Evolution." *The American Naturalist* 86 (January–February 1952): 33–44.

———. *Flowering Plants: Evolution above the Species Level.* Cambridge, Massachusetts: Harvard University Press, 1974.

———. *Processes in Organic Evolution.* Englewood Cliffs, New Jersey: Prentice Hall, 1977.

———. *The Ladyslipper and I.* Edited by Victoria C. Hollowell, Vassiliki Betty Smocovitis, and Eileen P. Duggan. St. Louis: Missouri Botanical Garden Press, 2007.

Stebbins, G. Ledyard, and Jack Major. "Endemism and Speciation in the California Flora." *Ecological Monographs* 35 (Winter 1965): 1–35.

Stebbins, G. Ledyard, and Dean William Taylor. *A Survey of the Natural History of the South Pacific Border Region.* Davis: University of California Institute of Ecology, 1973.

Stebbins, Robert C. *A Field Guide to the Western Reptiles and Amphibians.* Boston: Houghton Mifflin, 1966.

Steinbeck, John. *Novels and Stories: 1932–1937.* New York: Library of America, 1994.

Stock, Greg, Robert S. Anderson, and Robert C. Finkel. "Pace of Landscape Evolution of the Sierra Nevada, California, Revealed by Cosmogenic Dating of Cave Sediments." *Geology* 32 (March 2004): 193–196.

Sumner, F. B. "Some Biological Problems of Our Southwestern Deserts." *Ecology* 6 (1925): 352–371.

Thorne, Robert F. "A Historical Sketch of the Vegetation of the Mojave and Colorado Deserts of the American Southwest." *Annals of the Missouri Botanical Garden* 73 (1986): 642–651.

Van Dyke, Dix. *Daggett: Life in a Mojave Frontier Town.* Edited and with an introduction by Peter Wild. Baltimore, Maryland: Johns Hopkins University Press, 1997.

Van Dyke, John C. *The Desert: Further Studies in Natural Appearances.* Introduction by Peter Wild. Baltimore, Maryland: Johns Hopkins University Press, 1999.

Wells, Phillip V., and Rainer Berger, "Late Pleistocene History of Coniferous Woodland in the Mohave Desert." *Science* 155 (1966): 1640–1647.

Wells, Phillip V., and Juan H. Hunziker. "Origin of the Creosote Bush *(Larrea)* Deserts of Southwestern North America." *Annals of the Missouri Botanical Garden* 63 (1976): 843–861.

Wells, Phillip V., and Clive D. Jorgensen. "Pleistocene Wood Rat Middens and Climatic Change in Mohave Desert: A Record of Juniper Woodland." *Science* 143 (1964): 1171–1173.

Wernicke, Brian P., Robert Clayton, Mihai Ducea, Craig H. Jones, Stephen Park, Stan Ruppert, Jason Saleeby et al. "Origin of the High Mountains in the Continents: The Sierra Nevada." *Science* 271 (1996): 190–193.

Wild, Peter. "Sentimentalism in the American Southwest." In *Reading the West: New Essays on the Literature of the American West,* edited by Michael Kowalewski. Cambridge: Cambridge University Press, 1996.

———. *The Opal Desert: Explorations of Fantasy and Reality in the American Southwest.* Austin: University of Texas Press, 1999.

Wilson, Robert. *The Explorer King: Adventure, Science and the Great Diamond Hoax—Clarence King in the Old West.* New York: Scribner, 2006.

Wolfe, Jack A., Howard E. Schorn, Chris E. Forest, and Peter Molnar. "Paleobotanical Evidence for High Altitudes in Nevada During the Miocene." *Science* 276 (1997): 1672–1675.

Xenophon. *Anabasis.* Translated by Carleton L. Brownson. The Loeb Classical
Library. Cambridge, Massachusetts: Harvard University Press, 1998.
Zamudio, Graciela, and Gerardo Sanchez Diaz, editors. *Entre las Plantas y
la Historia: Homenaje a Jerzy Rzedowski.* Morelia, Michoacan: Universidad
Michoacana de San Nicolas de Hidalgo, 1998.

Index

Text   10/14 Palatino
Display   Univers and Bodoni
Compositor   BookMatters, Berkeley
Printer & Binder   Sheridan Books, Inc.